TUTORIAL CHEMISTRY TEXTS

D0206474

22

Organic Spectroscopic Analysis

ROSALEEN J. ANDERSON,
DAVID J. BENDELL &
PAUL W. GROUNDWATER

University of Sunderland

RS•C

ROYAL SOCIETY OF CHEMISTRY

ISBN 0-85404-476-0

A catalogue record for this book is available from the British Library

Published by The Royal Society of Chemistry, Thomas Graham House, Science Park, Milton Road, Cambridge CB4 0WF, UK
Registered Charity No. 207890
For further information see our web site at www.rsc.org

Typeset in Great Britain by Alden Bookset, Northampton
Printed and bound by Italy by Rotolito Lombarda

Preface

The aim of this textbook is not to provide you, the reader, with large numbers of correlation tables for every form of spectroscopy used by organic chemists; nor is it designed to give you an in-depth understanding of the physical processes that occur when molecules interact with electromagnetic radiation. What we hope to do in this textbook is to provide you with a basic understanding of how each type of spectroscopy gives rise to spectra, and how these spectra can be used to determine the structure of organic molecules. This text is aimed at undergraduate students in Years 1 and 2, and is meant to provide an introduction to organic spectroscopic analysis, leading to an appreciation of the information available from each form of spectroscopy and an ability to use spectroscopic information in the identification of organic compounds.

We will concentrate upon the most commonly used techniques in organic structure determination: nuclear magnetic resonance (NMR), infrared (IR) and ultraviolet-visible (UV-Vis) spectroscopy, and mass spectrometry (MS). The amount of space devoted to each technique in this text is meant to be representative of their current usage for structure determination.

Finally, we will try to bring all of these techniques together in an attempt to show you how to go about the structure determination of an unknown compound in a (reasonably) logical manner. Our aim has been to provide spectra to illustrate every point made, but do analyse fully each of the spectra in order to obtain the maximum information available. We hope you enjoy this text and find it useful in your studies.

R. J. Anderson, D. J. Bendell and P. W. Groundwater
Sunderland

TUTORIAL CHEMISTRY TEXTS

EDITOR-IN-CHIEF
Professor E W Abel

EXECUTIVE EDITORS
Professor A G Davies
Professor D Phillips
Professor J D Woollins

EDUCATIONAL CONSULTANT
Mr M Berry

This series of books consists of short, single-topic or modular texts, concentrating on the fundamental areas of chemistry taught in undergraduate science courses. Each book provides a concise account of the basic principles underlying a given subject, embodying an independent-learning philosophy and including worked examples. The one topic, one book approach ensures that the series is adaptable to chemistry courses across a variety of institutions.

TITLES IN THE SERIES

Stereochemistry *D G Morris*
Reactions and Characterization of Solids
  S E Dann
Main Group Chemistry *W Henderson*
d- and f-Block Chemistry *C J Jones*
Structure and Bonding *J Barrett*
Functional Group Chemistry *J R Hanson*
Organotransition Metal Chemistry *A F Hill*
Heterocyclic Chemistry *M Sainsbury*
Atomic Structure and Periodicity *J Barrett*
Thermodynamics and Statistical Mechanics
  J M Seddon and J D Gale
Basic Atomic and Molecular Spectroscopy
  J M Hollas
Organic Synthetic Methods *J R Hanson*
Aromatic Chemistry *J D Hepworth,*
  D R Waring and M J Waring
Quantum Mechanics for Chemists
  D O Hayward
Peptides and Proteins *S Doonan*
Biophysical Chemistry *A Cooper*
Natural Products: The Secondary
  Metabolites *J R Hanson*
Maths for Chemists, Volume I, Numbers,
  Functions and Calculus *M Cockett and*
  G Doggett
Maths for Chemists, Volume II, Power Series,
  Complex Numbers and Linear Algebra
  M Cockett and G Doggett
Nucleic Acids *S Doonan*

TITLES IN THE SERIES

Inorganic Chemistry in Aqueous Solution
  J Barrett
Organic Spectroscopic Analysis
  R J Anderson, D J Bendell and
  P W Groundwater

Further information about this series is available at www.rsc.org/tct

Order and enquiries should be sent to:
Sales and Customer Care, Royal Society of Chemistry, Thomas Graham House, Science Park, Milton Road, Cambridge CB4 0WF, UK

Tel: +44 1223 432360; Fax: +44 1223 426017; Email: sales@rsc.org

Contents

1

General Principles

Aims

This chapter introduces the interaction of electromagnetic radiation with organic molecules. By the end of the chapter you should be able to:

- Predict the region of the electromagnetic spectrum in which the different molecular transitions occur
- Understand the relationship between the energy of a transition and its frequency, wavelength and wavenumber
- Calculate the number of double bond equivalents in a molecule from its formula

1.1 The Interaction of Electromagnetic Radiation with Molecules

You will already know from your studies that the energy levels of atoms and molecules are quantized, *i.e.* there are discrete energy levels in atoms and molecules (Figure 1.1).

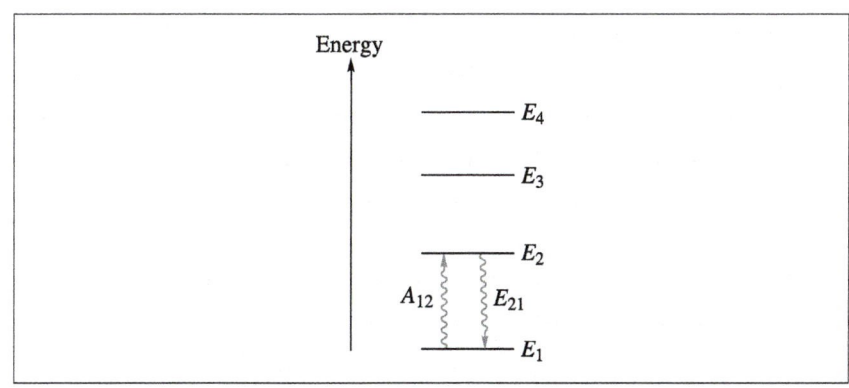

Figure 1.1 Discrete energy levels of an atom or molecule

In Figure 1.1, E_1 corresponds to the ground state of the atom/molecule, and E_2, E_3 and E_4 represent excited states. Given that these are all discrete energy levels, the difference in energy between them (and so the energy required for a particular transition) will also be a discrete value, which is called a quantum. We can see from Figure 1.1 that in order to excite the atom/molecule from E_1 to E_2 it must absorb an amount of energy equivalent to A_{12}. Excited states are generally short lived and relax back to the ground state by emission of energy, in this case E_{21}.

The energy difference between the excited state (E_2) and the ground state (E_1) will correspond to a certain frequency (v) or wavelength (λ) of electromagnetic radiation, and this will depend upon the type of transition (and hence the separation between energy levels). The relationship between the energy of a transition and the frequency is given by equation (1.1):

h is Planck's constant $(6.626 \times 10^{-34}$ J s).

$$\Delta E = hv \tag{1.1}$$

and so:

$c = \lambda v.$

$$\Delta E = hc/\lambda \quad \text{or} \quad \Delta E = hc\bar{v} \tag{1.2}$$

The energy of a particular transition is, therefore, proportional to the frequency or wavenumber ($\bar{v} = 1/\lambda$) and inversely proportional to the wavelength (equation 1.2).

The electromagnetic spectrum is divided into a number of regions. The names of these regions and the associated atomic/molecular transitions, together with the corresponding energies, frequencies, wavelengths and wavenumbers, are shown in Figure 1.2.

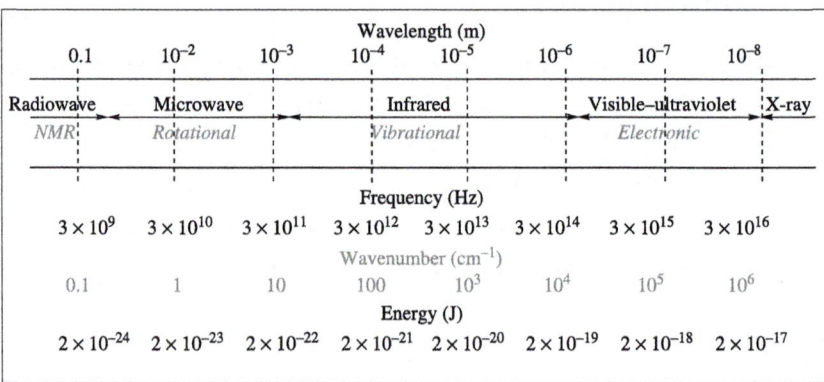

Figure 1.2 The electromagnetic spectrum

From Figure 1.2 we can see that nuclear magnetic resonance transitions (NMR), which correspond to wavelengths in the radiowave region of the spectrum, are those with the smallest gap between the energy levels, and electronic transitions in the ultraviolet-visible (UV-Vis) region have the largest energy gap between transition levels. The UV-Vis region is important, since absorptions in this region give rise to the colour associated with molecules; Figure 1.3 shows this region in more detail.

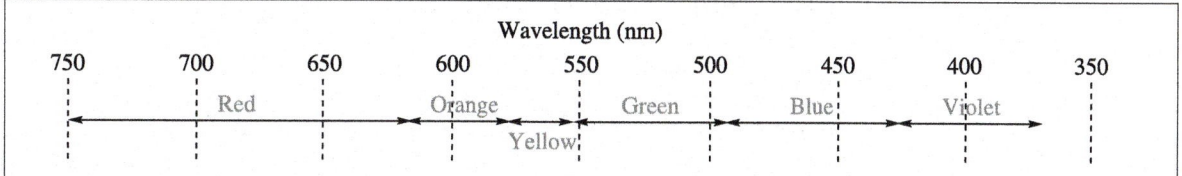

Figure 1.3   The visible region of the electromagnetic spectrum

X-rays have such high energy that they are capable of ionizing atoms and molecules, but they are also important in terms of structure determination using single-crystal X-ray diffraction. This technique is the ultimate in structure determination since it provides a "map" of the molecule in the crystal, *e.g.* Figure 1.4, but it is highly specialized, is limited to crystals and is not routinely available for all organic chemists, so it will not be discussed any further here.

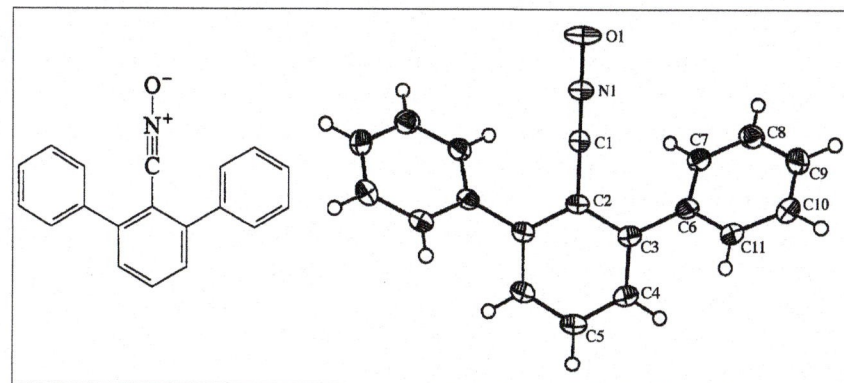

Figure 1.4   Single-crystal X-ray structure of 2,6-diphenylbenzo-nitrile oxide

Microwave (rotational) spectra are very complex, even for diatomic molecules, and give little useful information on organic molecules, which are relatively large. Rotational transitions are often responsible for the broadness of infrared (IR) bands, since each vibrational transition has a number of rotational transitions associated with it. The use of microwave spectroscopy is extremely rare in organic chemistry, and it too will be discussed no further here.

Transitions in all of the other forms of spectroscopy (UV-Vis, IR and NMR) are governed by selection rules that state which transitions are "allowed" and which are "forbidden" (although these latter transitions may still take place). We will mention the selection rules for each of these forms of spectroscopy when we study them in detail in Chapters 2–4.

Finally, one technique that does not rely upon electromagnetic radiation, mass spectrometry, will be discussed in detail in Chapter 5.

1.2   Double Bond Equivalents

Before we begin our study of these spectroscopic techniques, a reminder of a non-spectroscopic piece of information which is very useful in the determination of organic structures: the number of double bond equivalents, DBE (Ω), which tells us how many double bonds or rings are present in a molecule. Each double bond or ring reduces the number of hydrogens (or halogens) in a molecule by 2, so when we calculate the number of DBEs we simply compare the number of hydrogens which would be present in the fully saturated, acyclic compound with the number actually present, and divide by 2 to give the number of DBEs.

Box 1.1   Calculation of the Number of Double Bond Equivalents

For a neutral species, we can calculate the number of double bond equivalents, DBE (Ω), by comparing the molecular formula with that of the fully saturated, acyclic parent molecule with the same number of carbons and heteroatoms, $C_nH^*_{2n+2+y}N_yO_w$, where H^* is the total number of hydrogens and halogens.

To do this we can simply compare the number of hydrogens (and halogens) in the two formulae and, remembering that each double bond (or ring) results in a "loss" of 2 hydrogens from the molecular formula, calculate the number of DBEs. Alternatively, we can use the simple formula shown in equation (1.3):

$$\Omega = (C+1) - [(H^* - N)/2] \qquad (1.3)$$

where C = number of C atoms, H^* = number of hydrogens or halogens, and N = number of nitrogens.

For example, the molecular formula for 4-aminophenol (**1.1**) is C_6H_7NO. The formula of the fully saturated, acyclic parent structure [$C_nH^*_{2n+2+y}N_yO_w$, where H^* is the total number of hydrogens and halogens] would therefore be $C_6H_{(12+2+1)}NO$, *i.e.* $C_6H_{15}NO$. The difference in the number of hydrogens is therefore $15 - 7 = 8$, so there are 4 DBEs.

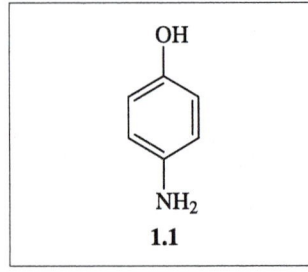

1.1

Alternatively, using equation (1.3):

$$\Omega = (6 + 1) - [(7 - 1)/2] = 7 - 3 = 4$$

Looking at the structure of 4-aminophenol, we can see that formally it has 3 double bonds, and the ring itself makes up the total of 4 DBEs.

Worked Problem 1.1

Q Calculate (a) the wavelength, (b) the frequency and (c) the energy associated with a transition with a wavenumber of 2260 cm^{-1}.

A (a) $\bar{v} = 1/\lambda$, so $\lambda = 1/\bar{v} = 1/2600 = 4.42 \times 10^{-4}$ cm

$$= 4.42 \times 10^{-6} \text{ m } (4.42 \text{ } \mu\text{m}).$$

(b) $c = v\lambda$, so $v = c/\lambda = 3 \times 10^8 \text{ ms}^{-1}/4.42 \times 10^{-6}$ m

$$= 6.79 \times 10^{13} \text{ s}^{-1}(6.79 \times 10^{13} \text{ Hz}).$$

(c) $E = hv = 6.626 \times 10^{-34} \text{ J s} \times 6.79 \times 10^{13} \text{ s}^{-1} = 4.5 \times 10^{-20} \text{ J}.$

Worked Problem 1.2

Q The local anaesthetic benzocaine has a molecular formula of $C_9H_{11}NO_2$. (a) Calculate the number of double bond equivalents; (b) identify all of the double bond equivalents in benzocaine (**1.2**).

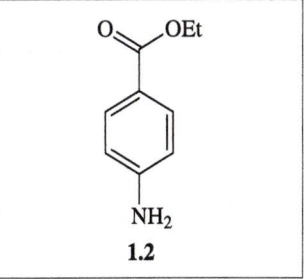

1.2

A (a) The formula of the fully saturated, acyclic parent structure [$C_nH^*_{2n+2+y}N_yO_w$, where H* is the total number of hydrogens and halogens] would therefore be $C_9H_{(18+2+1)}NO_2$, *i.e.* $C_9H_{21}NO_2$. The difference in the number of hydrogens is therefore $21-11=10$, corresponding to 5 DBEs. Alternatively, $\Omega=(9+1)-[(11-1)/2]= 10-5=5$.

(b) Benzene ring $\equiv$ 4 DBEs ($3 \times C=C$ and 1 ring) and $C=O$.

Summary of Key Points

1.   The energy levels of atoms and molecules (and so the difference in energy between these levels) have discrete values (quanta).

2.   The wavenumber of a transition is inversely proportional to the wavelength ($\bar{v} = 1/\lambda$).

3.   The relationship between the energy of a transition and the frequency is given by $\Delta E = hv$ or $\Delta E = hc/\lambda$ or $\Delta E = hc\bar{v}$, where h is Planck's constant. The energy of a particular transition is, therefore, proportional to the frequency or wavenumber, and inversely proportional to the wavelength.

4.   NMR transitions correspond to wavelengths in the radiowave region of the spectrum, vibrational transitions correspond to wavelengths in the IR region, and electronic transitions to the UV-Vis region.

5.   The number of double bond equivalents corresponds to the difference between the molecular formula and that for the saturated acyclic parent compound. Each DBE (double bond or ring) results in the subtraction of 2 hydrogens or halogens from the molecular formula of this parent structure.

Problems

1.1.   Calculate the energy associated with transitions with the following frequencies, wavelength or wavenumber. What type of molecular transition is associated with each transition? (a) $v = 3 \times 10^8$ Hz; (b) $\lambda = 254$ nm (254×10^{-9} m); (c) $\lambda = 1.0$ cm; (d) $\bar{v} = 2600$ cm^{-1}; (e) $v = 4.1 \times 10^{14}$ Hz; (f) $\bar{v} = 2 \times 10^8$ m^{-1}.

1.2.   How many double bond equivalents are there in each of the following molecules: (a) $C_6H_{12}O_6$; (b) $C_8H_9NO_2$; (c) $C_{27}H_{46}O$; (d) $C_5H_5N_5$; (e) $C_6H_3Cl_3O$; (f) $C_{46}H_{58}N_4O_9$.

2
Ultraviolet–Visible (UV-Vis) Spectroscopy

Aims

This chapter describes how ultraviolet–visible (UV-Vis) spectroscopy is used in organic chemistry. After you have studied this chapter, you should be able to:

- Describe briefly how the absorbance of a UV-absorbing compound is obtained
- Explain the relationship between the absorption wavelength and the energy difference between the energy levels involved
- Calculate the molar absorptivity of a molecule from its absorbance, concentration and the path length
- Recognize chromophores, and explain how conjugation and aromatic substituents can contribute to chromophores
- Explain how pH can affect the chromophore and the UV-Vis spectra of acidic and basic compounds
- Predict the absorbance maximum for unsaturated compounds using the Woodward–Fieser rules
- Summarize some practical applications of UV-Vis spectroscopy

2.1 Instrumentation

There are several different types of UV-Vis spectrometers, with the usual instrument met in an undergraduate laboratory being a double (or dual) beam spectrophotometer, which consists of a UV-visible light source, two cells through which the light passes, and a detector (usually a photomultiplier) to measure the amount of light passing through the cells. There are basic spectrometers that measure the absorbance at a specific wavelength, set by the user, and others that can scan the entire

UV-Vis range. Newer spectrometers are usually computer controlled and allow the user greater flexibility, *e.g.* in overlaying spectra of a reaction mixture over time, or constructing a calibration graph to determine the concentration of an unknown.

Single-beam UV-Vis spectrometers work on the same general principles, but measure the absorption of the reference first, followed by the sample. They can scan across the entire UV-Vis range or can be used at a single wavelength. Detector technology has improved recently: the diode array detector enables simultaneous detection over the entire range to be achieved, allowing rapid quantification of absorbing species.

The term UV-Vis normally applies to radiation with a wavelength in the range 200–800 nm. There are many groups that absorb below 200 nm, but this part of the spectrum is difficult to examine (as oxygen absorbs UV radiation below 200 nm) unless the spectra are recorded in a vacuum (vacuum UV-Vis).

In the double-beam UV-Vis spectrophotometer the light is split into two parallel beams, each of which passes through a cell; one cell contains the sample dissolved in solvent and the other cell contains the solvent alone. The detector measures the intensity of the light transmitted through the solvent alone (I_0) and compares it to the intensity of light transmitted through the sample cell (I). The absorbance, A, is then calculated from the relationship shown in equation (2.1):

$$A = \log_{10} \frac{I_0}{I} \qquad (2.1)$$

Ethanol is transparent to UV above 200 nm and is, therefore, commonly used as the solvent in UV-Vis spectroscopy.

2.2   Selection Rules and the Beer–Lambert Law

We can calculate the energy of a particular wavelength using equation (2.2):

$$E(\text{kJ mol}^{-1}) = \frac{1.19 \times 10^5}{\lambda(\text{nm})} \qquad (2.2)$$

As the relationship between E and λ is a reciprocal one, we can see that short wavelength radiation corresponds to high energy and long wavelengths correspond to low energy.

As you will have seen in Chapter 1, light at the short wavelength end of the electromagnetic spectrum has enough energy to promote electronic transitions in organic molecules, such that absorption of UV light (200–400 nm; 595–299 kJ mol^{-1}) or Vis light (400–800 nm; 299–149 kJ mol^{-1}) can result in the promotion of outer electrons from one electronic energy level to a higher one.

The difference between electronic energy levels is greater than the difference between any other molecular energy levels, so these transitions require the higher energy of short wavelength radiation. Promotion between other energy levels, *i.e.* vibrational or rotational, requires only the lower energy infrared (vibrational) or microwave (rotational) radiation (Figure 2.1).

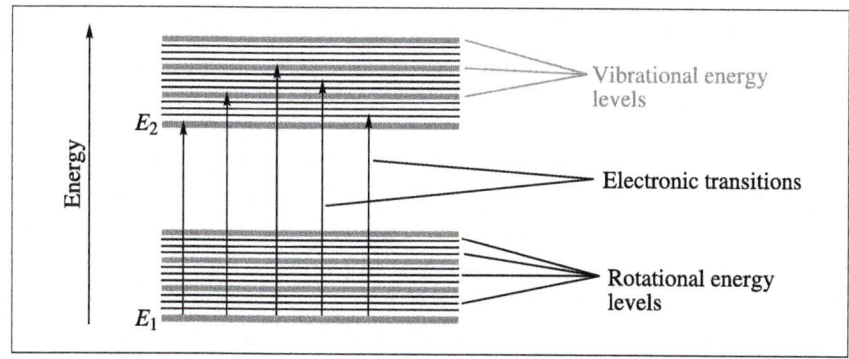

Figure 2.1 Schematic diagram showing possible molecular electronic transitions, and vibrational and rotational energy levels

At room temperature, the majority of molecules are in the lowest vibrational state of the lowest electronic energy level, the "ground state" (E_1). Absorption of UV or visible light leads to promotion of an electron from E_1 to the excited electronic level, E_2. The electronic transition is accompanied by vibrational and rotational transitions, so that the promotion of an electron can occur from the ground state electronic energy level of E_1 to any of the vibrational or rotational energy levels of E_2. This explains why UV-Vis absorption bands are characteristically broad, although energy transitions between rotational and vibrational energy levels within one electronic energy level can show fine structure (*e.g.* see Chapter 3 for many examples of sharp absorbances). Fine structure in UV spectra can sometimes be achieved by using a solvent that has little interaction with the sample molecules.

The fact that there are many electronic transitions possible, however, does not mean that they can or will occur. There are complex selection rules based on the symmetry of the ground and excited states of the molecule under examination. Basically, electronic transitions are allowed if the orientation of the electron spin does not change during the transition and if the symmetry of the initial and final functions is different; these are called the spin and symmetry selection rules, respectively. However, the so-called "forbidden" transitions can still occur, but give rise to weak absorptions.

According to the Beer–Lambert law, the absorbance, A, of a solution is directly proportional to the path length (l, length of the cell containing the solution, in cm) and the concentration of the absorbing molecule (c, in moles per litre), according to equation (2.3):

$$A = \varepsilon c l \qquad (2.3)$$

where $\varepsilon =$ the molar absorptivity of the absorbing molecule, sometimes called the molar extinction coefficient, and is characteristic of the molecule. The molar absorptivity is usually expressed in units of $100 \text{ cm}^2 \text{ mol}^{-1}$, although it is becoming more common to use units of $\text{dm}^3 \text{ mol}^{-1} \text{ cm}^{-1}$ (these units give values of ε that differ by a factor of n denotes non-bonding and * an anti-bonding orbital (excited state).

10 from those values in units of 100 cm^2 mol^{-1}). The molar absorptivity, ε, is a measure of the intensity of the absorption and usually ranges from 0 to 10^6 (units of 100 cm^2 mol^{-1}). The greater the probability of a particular absorption and its associated electronic transition, the greater the ε value for that transition. For most molecules, absorptions associated with $\pi \rightarrow \pi^*$ transitions have higher ε values than the $n \rightarrow \pi^*$ transitions.

In general, forbidden transitions give rise to low-intensity (low ε) absorption bands ($\varepsilon < 10,000$), but two important "forbidden" absorptions are seen quite commonly: the $n \rightarrow \pi^*$ transition of ketones at approximately 300 nm (ε usually 10–100), and the weak $\pi \rightarrow \pi^*$ absorption of benzene rings at about 260 nm (ε about 100–1000).

2.3   Chromophores

The part of the molecule containing the electrons involved in the electronic transition which gives rise to an absorption is called the chromophore. The wavelength of the maximum of the broad absorption is labelled λ_{max}. Most of the simple, non-conjugated chromophores give rise to high-energy (low-wavelength) absorptions with low-intensity (ε) values, but the majority of these absorptions are lost in atmospheric oxygen absorptions (Table 2.1).

Table 2.1   Absorption wavelengths of simple, non-conjugated chromophores

λ_{max}/nm	Chromophore	Transition causing absorption
~150	C—C or C—H σ-bonded electrons	$\sigma \rightarrow \sigma^*$
~185–195	—X: (X = O, N, S) Lone pair electrons	$n \rightarrow \sigma^*$
~300	C=O:	$n \rightarrow \pi^*$
~190	Lone pair electrons	$n \rightarrow \sigma^*$
~190	C=C (isolated) π-bonded electrons	$\pi \rightarrow \pi^*$

With the exception of the high-energy $\sigma \rightarrow \sigma^*$ transition of saturated alkyl systems, these transitions require either a lone pair or a π-bond from which the electron can be promoted. In fact, UV spectra are generally only of interest if the system is unsaturated; chromophores with the greatest degree of unsaturation give rise to the most intense absorptions at longest wavelength.

We can show electronic transitions using molecular orbital diagrams, *e.g.* the $\pi \rightarrow \pi^*$ transition of ethene, CH$_2$=CH$_2$, is shown in Figure 2.2. Here we can see that UV radiation at 190 nm provides the required energy (626.3 kJ mol^{-1}) to promote a bonding electron from the π-bonding orbital (the highest occupied molecular orbital, HOMO) to the π*-antibonding orbital (the lowest unoccupied molecular orbital, LUMO), resulting in the excited state of ethene, in which the π-electrons are unpaired and the HOMO is now the π*-antibonding orbital.

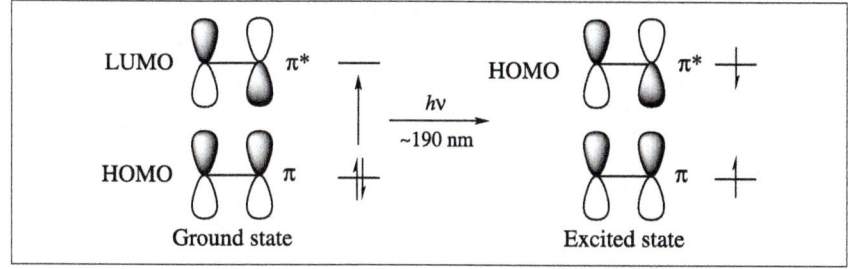

Figure 2.2 The $\pi \rightarrow \pi^*$ transition of ethene

As mentioned previously, the intensity of the major absorption, and of course the ε value, increases as the chromophore increases in length. As conjugation increases the chromophore, *systems with a higher degree of conjugation have greater intensity absorption bands with larger ε values.* Increased conjugation decreases the energy difference between the HOMO and the LUMO of a particular system, which means that less energy is required to promote an electron to a higher energy level. As lower energy transitions appear at higher wavelengths, it follows that, in general, *the longer the chromophore, the longer the wavelength of the absorption maximum.*

The effect of conjugation on the UV-Vis spectrum of a simple chromophore is shown in Table 2.2, in which we can see that the conjugated MVK has an increased λ_{max} and ε for both absorption bands.

Table 2.2 Effect of conjugation on a simple chromophore

Transition	Propanone (acetone) $\underset{Me}{\overset{O}{\underset{\qquad}{\parallel}}}\underset{\qquad}{\overset{\quad}{C}}Me$ λ_{max} (ε)	But-3-en-2-one (methyl vinyl ketone, MVK) $\underset{Me}{\overset{O}{\underset{\qquad}{\parallel}}}\underset{\qquad}{\overset{\quad}{C}}CH{=}CH_2$ λ_{max} (ε)
$\pi \rightarrow \pi^*$ (C=C)	187 nm (900)	219 nm (3600)
$n \rightarrow \pi^*$ (C=O)	270 nm (15)	324 nm (24)

An intense absorption above 210 nm indicates the presence of a conjugated system, and the longer the wavelength, the longer the chromophore.

β-Carotene (Figures 2.3 and 2.4) is the classic example used to illustrate how a lengthened chromophore produces an absorption at longer wavelength. β-Carotene has 11 conjugated π-bonds, which leads to a much smaller energy difference between the π and π* molecular orbitals than for an isolated double bond. This translates into a series of weak absorptions up to 500 nm: in the visible region.

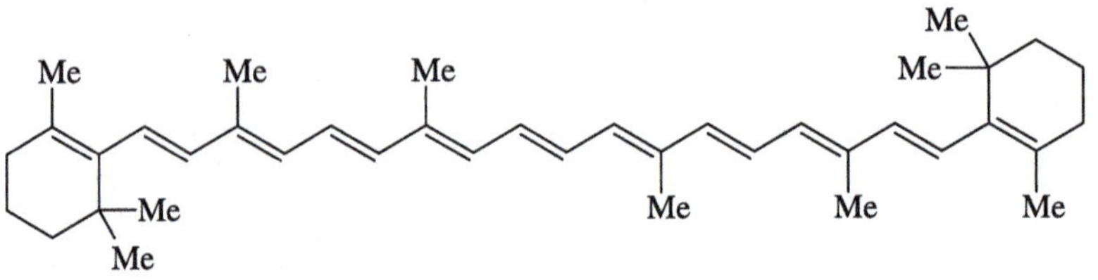

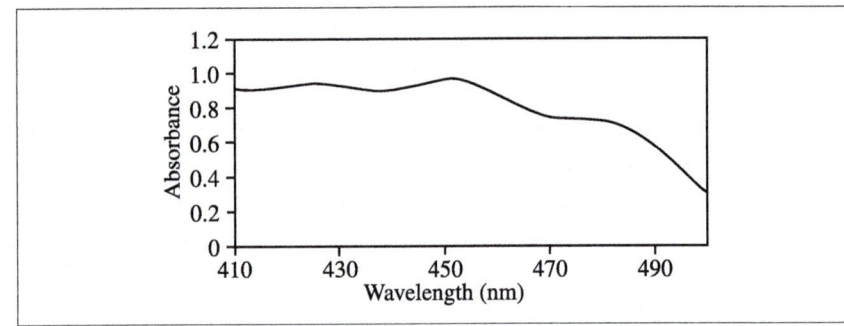

Figure 2.3 The orange-coloured pigment β-carotene (λ_{max} 426, 451 and 483 nm), found in carrots

Figure 2.4 The UV spectrum of β-carotene (5.8×10^{-5} mol dm^{-3} in ethanol) at high wavelength

When white light (light of all wavelengths) is passed through a solution of a compound which absorbs in the visible region, the compound absorbs light of the appropriate wavelength (energy) to promote an electron and reflects the remaining light. The colour we perceive is the reflected light, and is complementary in colour to the light that was absorbed. Table 2.3 shows the relationship between absorbed and reflected (perceived) colour and wavelength.

Table 2.3 The relationship between absorbed and reflected (perceived) colour

Perceived colour of compound	Colour absorbed by compound	Wavelength of absorbed light (nm)
Green-yellow	Violet	400–424
Yellow	Blue	424–491
Red	Green	491–570
Blue	Yellow	570–585
Green–blue	Orange	585–647
Green	Red	647–700

After absorption of the blue–green colour from white light (carotene absorbs light of wavelength up to about 500 nm), the complementary colour (red–yellow) remains to be reflected and perceived by our eyes. β-Carotene, therefore, appears intensely orange coloured; as the name implies, it is found in high concentrations in carrots.

The chromophore is not only extended by conjugation: the presence of an electron-donating substituent on an aromatic ring also has the effect of increasing the wavelength of absorption, as we can see for the examples in Box 2.1.

An increase in the λ_{max}, towards the red end of the absorbed spectrum, is called a red shift (shift to longer λ, lower energy) or a **bathochromic shift**; the greater the red shift, the more blue or blue–green the colour perceived (remember, the absorbed and perceived colours are complementary). The opposite effect is a shift to shorter wavelength, towards the blue end of the absorbed spectrum (higher energy) and is called a blue shift or a **hypsochromic shift**; colour associated with these wavelengths is more red or yellow to our eyes.

Box 2.1 Effect of Conjugation and Electron-donating Substituents on Wavelength

Name	Benzene	Phenol	Benzoic acid	Aniline
		:ÖH	O=C–OH	:NH₂
Solvent	Cyclohexane	Ethanol	Methanol	Water
λ_{max}/nm (ε) (ethanol)	203 (7500)	220 (6150)	230 (11,900)	230 (8600)
	254 (210)	273 (1900)	275 (900)	280 (1430)
Effect of substituent	–	Electron donating, conjugated	Electron withdrawing, conjugated	Electron donating, conjugated

Notice how substituents with lone pairs on the atom attached to the aromatic ring (*e.g.* OH, NH$_2$) extend the chromophore by resonance of the lone pair electrons into the aromatic ring and, in general, lead to an increase in both λ_{max} and ε, and have a similar effect to that of conjugation (*e.g.* CO$_2$H).

2.3.1 Role of the Solvent

We have seen how the position and intensity of λ_{max} is affected by the energy difference between the electronic energy levels. The Franck–Condon principle states that electronic transitions involve the movement of electrons, including those of the solvent, but not the movement of atoms. When the solvent electrons can rearrange to stabilize the excited state of a molecule, the energy difference between the electronic levels of the molecule is lowered and the absorption moves to higher wavelength.

To help us understand how a polar solvent can help to stabilize an excited state, we will consider the $\pi \to \pi^*$ transition of an alkene. We can represent the ground state and excited state species in a simple way with resonance structures. It is important to realize, though, that the dipolar structures in Scheme 2.1 are not the excited state but they do make a (minor) contribution to the excited state structure.

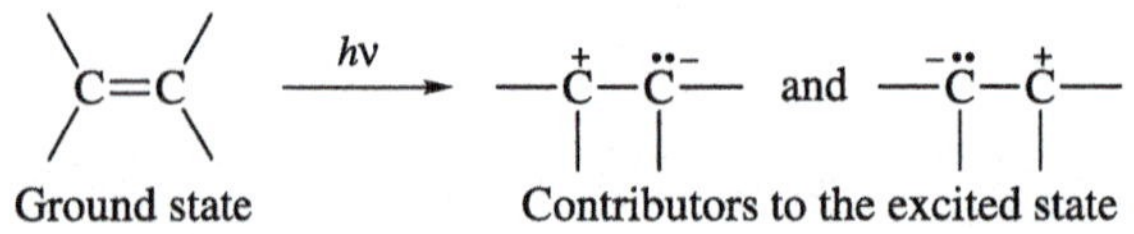

Scheme 2.1 Resonance representations of the ground and excited states of an alkene

Even with our simplistic example, we can see that a polar solvent, such as ethanol, will stabilize the dipolar species on the right-hand side of the equation better than a non-polar solvent, such as cyclohexane. This extra stabilization of the excited state (LUMO) by ethanol leads to a decrease in the energy difference between the HOMO and LUMO and a shift to higher wavelength, a red (bathochromic) shift, when the spectrum is recorded in ethanol.

Conversely, in the case of the $n \to \pi^*$ transition of a ketone, the interaction of the lone pair of the ground state carbonyl with a polar solvent lowers the energy of the lone pair n orbital, and thereby increases the energy required to promote an electron to the π^* energy level. Consequently, there is often a blue (hypsochromic) shift – to lower wavelength – observed when the UV spectrum of a ketone is recorded in ethanol rather than cyclohexane.

2.3.2 Effect of pH

Dramatic changes to the UV-Vis spectra of some compounds, especially certain substituted aromatic compounds, occur with a change in the pH of the solvent.

Phenols and substituted phenols are acidic and display striking changes to their absorptions upon the addition of base. Removal of the acidic phenolic proton increases the conjugation of the lone pairs on the oxygen with the π-system of the aromatic ring, as shown in Scheme 2.2, leading to a decrease in the energy difference between the HOMO and LUMO orbitals, and an associated red or bathochromic shift (to longer wavelength), along with an increase in the intensity of the absorption (Figure 2.5).

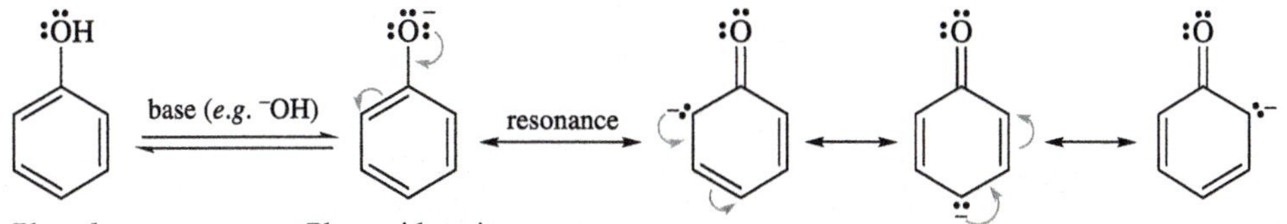

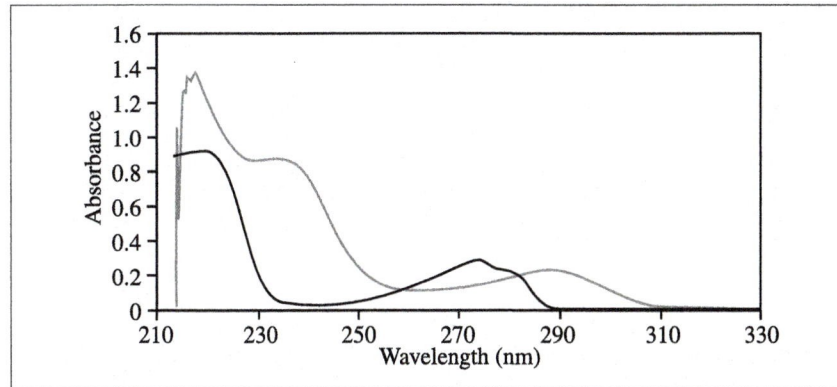

Phenol

Phenoxide anion

Neutral or acidic conditions

λ_{max} (H₂O) 210 (6000)
270 (1500)

Basic conditions

λ_{max} (H₂O) 235 (9400)
287 (2600)

Scheme 2.2 Effect of pH upon the absorption maxima of phenols

Figure 2.5 The UV-Vis spectra of phenol (0.15 mmol dm^{-3} in ethanol, black) and the phenolate ion (0.074 mmol dm^{-3} in 0.1 M NaOH, red)

Treatment of an aromatic amine with acid, on the other hand, causes protonation of the amine (Scheme 2.3), and this leads to a loss of the overlap between the amine lone pair and the aromatic π-system. The result is a blue or hypsochromic shift (to shorter wavelength) along with a decrease in intensity (Figure 2.6).

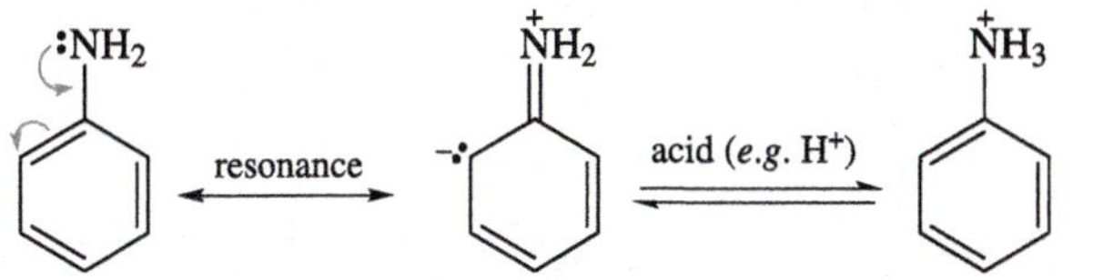

Aniline

Protonated

Neutral or basic conditions

λ_{max} (H₂O) 235 (14800)
285 (2800)

Acidic conditions

λ_{max} (H₂O) 203 (7500)
254 (180)

Scheme 2.3 Effect of pH upon the absorption maxima of an aromatic amine

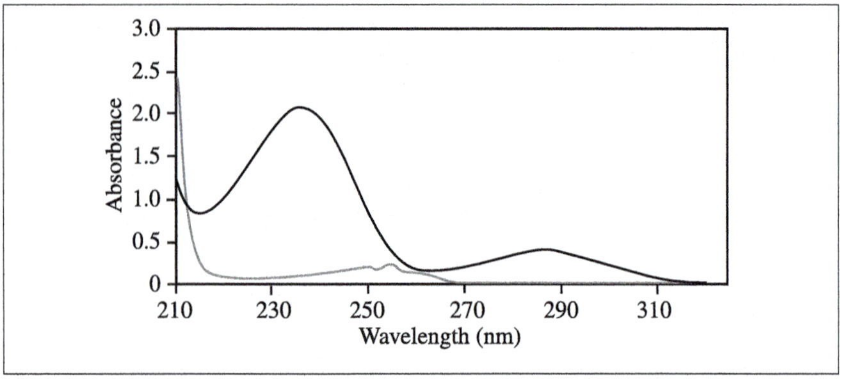

Figure 2.6 UV-Vis spectra of aniline (0.14 mmol dm^{-3} in ethanol, black) and the anilinium ion (1.4 mmol dm^{-3} in 0.1 M HCl, red)

The pK_a (= $-\log_{10}K_a$) is a measure of acid strength, with the lower the value the stronger the acid. It is the pH at which a weak acid is 50% ionized:

$$HA \rightleftharpoons H^+ + A^-$$

$$K_a = \frac{[H^+][A^-]}{[HA]}$$

Many pH indicators owe their utility to their absorptions in the visible region of the UV-Vis spectrum. Changes to the pH lead to changes in the indicator chromophore, and result in reliable colour changes at predictable pH values. One such example is that of phenolphthalein, which is a phenol and can be deprotonated at elevated pH to give the anion (Scheme 2.4), extending the chromophore and leading to a

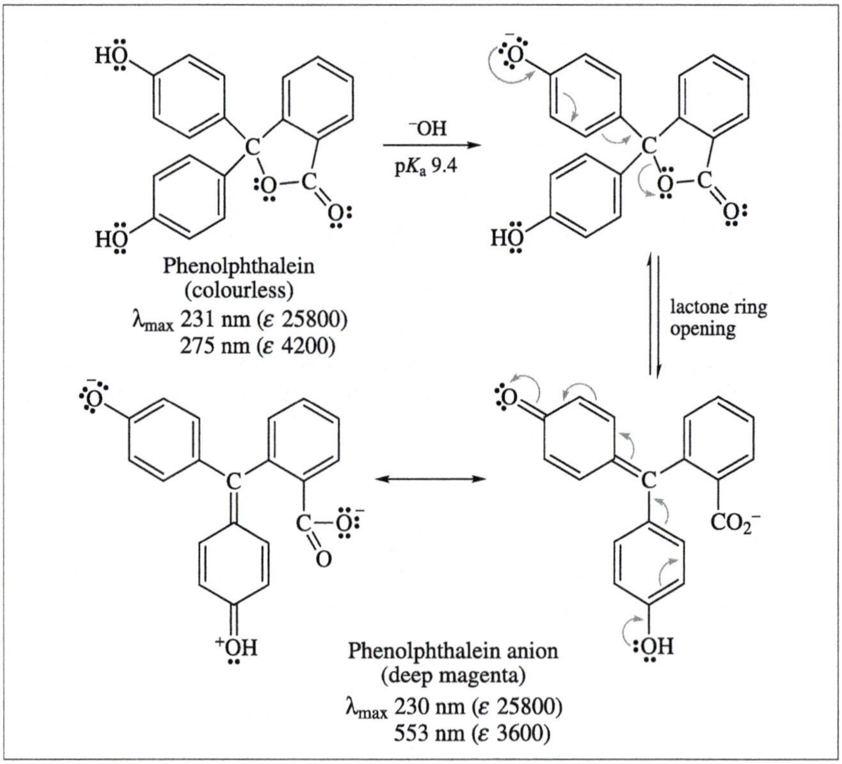

Phenolphthalein
(colourless)
λ_{max} 231 nm (ε 25800)
275 nm (ε 4200)

lactone ring opening

Phenolphthalein anion
(deep magenta)
λ_{max} 230 nm (ε 25800)
553 nm (ε 3600)

Scheme 2.4 Acid–base equilibria of phenolphthalein: colourless to magenta (only the major resonance contributors to the anion are shown) substantial bathochromic shift (to longer wavelength). Thus the anion of phenolphthalein is deep magenta in colour, while un-ionized phenolphthalein is colourless (Figure 2.7). The pK_a of this acid–base equilibrium is 9.4: at acidic and neutral pH there is insufficient anion to detect the colour by eye, and it appears colourless. As the pH approaches the pK_a, the concentration of anion increases and, at pH 8.2, the colour becomes visible to the eye. As pH 8.2 is close to neutrality, phenolphthalein is widely used to show the end-point in weak acid–strong base titrations.

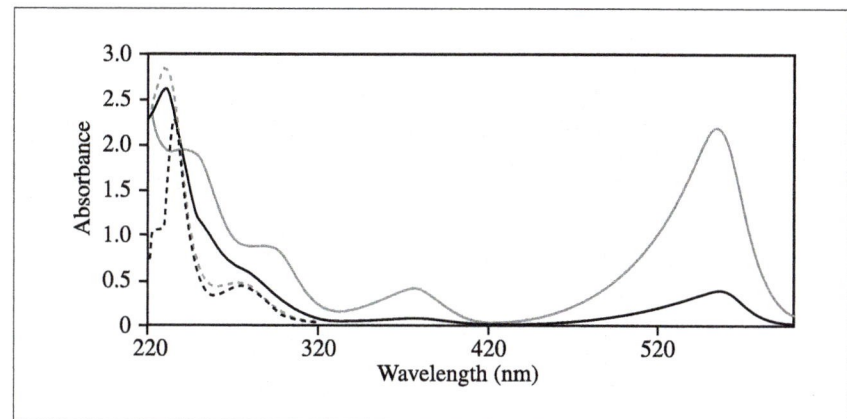

Figure 2.7 The UV-Vis spectra of phenolphthalein (0.103 mmol dm^{-3}) at pH 13.0 (upper red line), pH 9.0 (upper black line), pH 8.0 (dashed red line) and pH 4.0 (dashed black line)

2.3.3 Woodward–Fieser Rules

As we have already seen, delocalization of electrons by conjugation decreases the energy difference between the HOMO and LUMO energy levels, and this leads to a red shift. Alkyl substitution on a conjugated system also leads to a (smaller) red shift, due to the small interaction between the σ-bonded electrons of the alkyl group with the π-bond system. These effects are additive, and the empirical Woodward–Fieser rules were developed to predict the λ_{max} values for dienes (and trienes). Similar sets of rules can be used to predict the λ_{max} values for α,β-unsaturated aldehydes and ketones (enones) and the λ_{max} values for aromatic carbonyl compounds. These rules are summarized in Table 2.4.

Table 2.4 UV absorption wavelength correlations for conjugated systems

	Dienes	Enones	Aromatic carbonyls

1. Use the appropriate parent value

Dienes	Enones	Aromatic carbonyls
215 nm	X = H 207 nm	X = H 250 nm
254 nm (when homocyclic diene – within one ring)	X = C 215 nm (acyclic or part of six-membered ring)	X = C 246 nm
	X = C 202 nm (part of five-membered ring)	X = OH, OR 230 nm

2. Add an increment for any extra conjugated π-bonds

Dienes		Enones	Aromatic carbonyls
C=C	30	30	–
C=C–C=C	40	40	–
Exocyclic C=C	5	5	–

3. Add increments for any substituents

Substituent	Dienes	Enones α	Enones β	Enones γ	Enones δ	Aromatic 2-	Aromatic 3-	Aromatic 4-
H	0		0				0	
R (CH₃, etc.)	5	10	12	18	18	3	3	7
Cl	5	15	12	–	–	0	0	10
OH, OR	5	35	30	–	–	7	7	25
OCOR	0	6	6	–	6	–	–	–
O⁻	–	–	75	–	–	15	15	80
SR	30	–	85	–	–	–	–	–
NR₂	60	–	95	–	–	20	20	85

Using the Woodward–Fieser rules, we can predict the absorption maximum of an unsaturated compound with pleasing accuracy, generally to within ± 5 nm of the observed value. For example, the calculated absorbance maximum of ergocalciferol (vitamin D_2) of 265 nm (Figure 2.8) is only 1 nm different from the observed value at 266 nm (Figure 2.9).

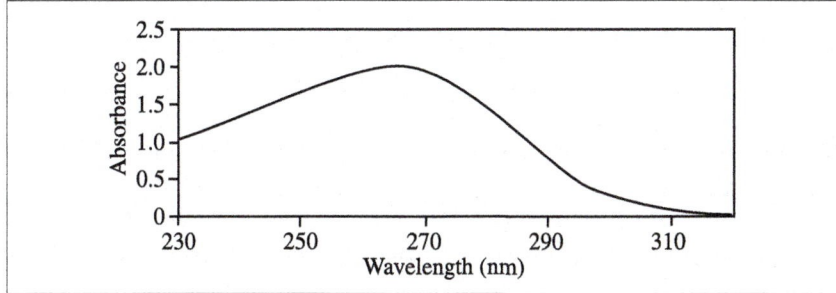

Calculated λ_{max}	nm
Parent diene	215
Extra C=C (conjugated)	30
Exocyclic position	5
C substituent × 3	15
	$\underline{265}$

Figure 2.8 Ergocalciferol (vitamin D_2) and the calculated absorption maximum

Figure 2.9 UV-Vis spectrum of ergocalciferol (0.12 mmol dm^{-3} in ethanol)

2.4 Applications of UV Spectroscopy

Before NMR spectroscopy and mass spectrometry revolutionized the structural elucidation of organic molecules, UV spectroscopy was an important technique and was used to identify the key chromophore of an unknown molecule. The importance of UV is much diminished nowadays, but it still retains its place in certain applications, such as the determination of kinetic parameters, K_M (the Michaelis constant) and k_{cat} (the turnover rate of an enzyme, in molecules per second), for a number of enzymic reactions and in the analysis of pharmaceuticals.

2.4.1 UV in the Analysis of Pharmaceutical Preparations

The molar absorptivity (ε) of a known molecule is constant under identical conditions of solvent, concentration and path length, and can be used to quantify the amount of a particular pharmaceutical in a tablet. Such assays form the basis of many quality assurance procedures in the pharmaceutical industry, and have been extensively used by the British Pharmacopoeia (B.P.). More recently, however, high-performance liquid chromatography (HPLC) has replaced UV analysis in many B.P. assays, as most industrial analyses routinely use HPLC.

In the pharmaceutical industry, medicines are standardized to a particular weight of active pharmaceutical per unit (for example, mg of substance per tablet), so the weight of a substance in a tablet, or volume of medicine, is of more interest than the number of moles. For most pharmaceutical preparations, then, we use the specific absorbance, $A(1\%, 1\text{ cm})$, which is the absorbance ($\log_{10} I_0/I$) of a 1% w/v solution (*i.e.* 1 g of substance in 100 cm^3 solvent) in a 1 cm path length cell, in place of the molar absorptivity, ε. This is the absolute method of substance identification, and the absorbance, A, can be related to $A(1\%, 1\text{ cm})$ by equation (2.4):

$$A = A(1\%, 1\text{ cm}) \times \text{concentration} \times \text{path length} \qquad (2.4)$$

The absolute method relies on the UV spectrophotometer being accurate in the measurement of wavelength and intensity, and the B.P. specifies methods for the calibration of both. However, UV-Vis spectra are not specific for any one particular substance, as many UV absorbing molecules have similar UV spectra, and this method is used in conjunction with several identification tests, as described in the B.P.

Derivatization: to prepare a derivative of a compound (*e.g.* an ester derivative of a carboxylic acid).

A number of B.P. tests require derivatization of the substance under investigation before the UV absorbance is recorded. In these cases, in order to take account of the possibility of the reaction not proceeding to completion, the comparative method is employed, which makes use of a reference (defined in the B.P.). The concentrations and absorbances of the reference and substance under investigation are related by equation (2.5):

$$\frac{C_1}{C_2} = \frac{A_1}{A_2} \qquad (2.5)$$

2.4.2 UV Detection in Chromatography

UV spectroscopy is exploited for the detection of analytes in a number of separative analytical techniques, such as thin layer chromatography (TLC) and HPLC.

Thin Layer Chromatography

TLC is a simple technique that is commonly used for the rapid qualitative analysis of reaction mixtures. Many organic compounds used in chemical and pharmaceutical manufacture, and in organic and medicinal chemistry research, contain an aromatic ring or other UV-absorbing system which can be visualized on the TLC plate under a UV lamp.

The whole procedure takes about 5 minutes and is a much-valued technique in the laboratory.

UV Detectors in HPLC

HPLC has transformed quality control in the pharmaceutical and chemical industries, as it provides a rapid means of checking the purity of samples and even allows for the purification of small amounts of samples by preparative HPLC. The majority of such systems use UV to detect and quantify substances as they elute from the separative column. UV detectors are usually variable wavelength and can be used to detect molecules with absorption maxima above 210 nm by measuring the absorbance of the eluent. When a UV-absorbing substance is eluted from the HPLC column, it absorbs UV radiation at the appropriate wavelength for its chromophore. The amount of UV absorbed is proportional to the quantity of substance being eluted, and is converted into a peak on a chart recorder. Integration of each peak allows the relative quantities of the components of the solute to be determined.

Summary of Key Points

1. UV-Vis spectroscopy involves the promotion of electrons from bonding or non-bonding orbitals to anti-bonding orbitals.

2. The UV spectrum arises from absorption of UV or visible light of the appropriate energy for a particular electronic transition. The absorbance (A) at any wavelength is calculated from the intensity of light transmitted through a solution of the sample in solvent (I) compared to the intensity of light transmitted through the solvent alone (I_0): $A = \log_{10}(I_0/I)$.

3. The molar absorptivity (ε) of an absorption maximum (λ_{max}) gives an indication of the probability of that particular electronic transition and is characteristic for a given molecule. It can be calculated using Beer–Lambert's law to relate absorbance (A) to the molar absorptivity (ε), the concentration (c, mol L^{-1}) and the path length (l, cm): $A = \varepsilon c l$.

4. The most useful UV absorbances are those of conjugated organic molecules, which arise from non-bonding and π-orbitals. Systems with a higher degree of conjugation have absorption bands with increased intensities and larger ε values; the longer the chromophore, the longer the wavelength of the absorption maximum.

5. The chromophore of aromatic systems is increased by conjugation with a substituent, *e.g.* those with π-electrons or lone pairs of electrons, in a predictable manner.

6. The absorbance maximum of phenols is increased by the addition of base and formation of the phenolate ion, with an associated absorbance shift to longer wavelength (a bathochromic or red shift), whilst the absorbance maximum of anilines is decreased by the addition of acid, which causes protonation of the amine and loss of the lone pair overlap with the π-system, leading to an absorbance shift to shorter wavelength (a hypsochromic or blue shift).

7. Using the Woodward–Fieser rules, the λ_{max} of dienes and trienes, α,β-unsaturated aldehydes and ketones (enones), and for aromatic carbonyl compounds can be reliably calculated and predicted.

Problems

2.1. The UV spectra of aniline and 4-nitroaniline are shown in Figure 2.10. Account for the red shift seen in the spectrum of 4-nitroaniline when compared to that of aniline.

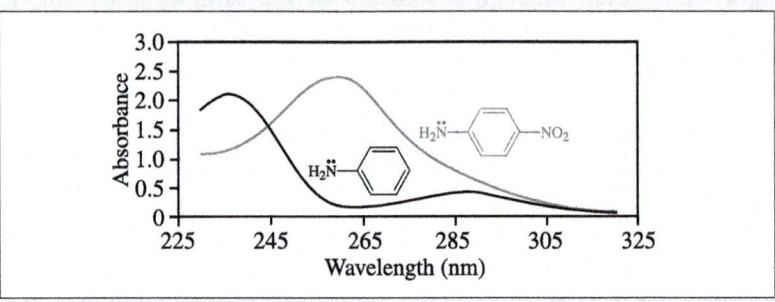

Figure 2.10 UV-Vis spectra of aniline (0.14 mmol dm^{-3} in ethanol, black) and 4-nitroaniline (0.14 mmol dm^{-3} in ethanol, red)

2.2. Calculate the expected λ_{max} for carvone and piperonal using the Woodward–Fieser rules. Compare your values to the observed λ_{max} in the spectra in Figure 2.11.

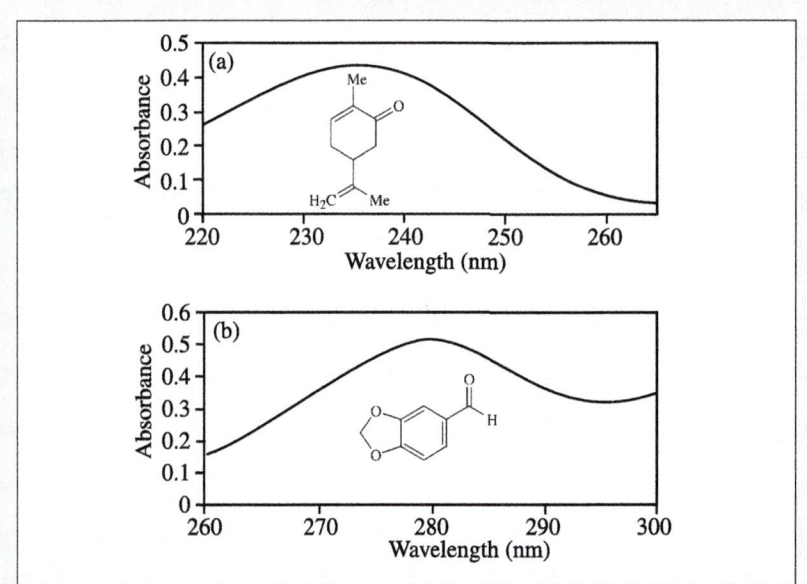

Figure 2.11 UV spectra of (a) carvone (0.043 mmol dm^{-3} in ethanol, $A = 0.432$) and (b) piperonal (0.067 mmol dm^{-3} in ethanol, $A = 0.509$)

2.3. Calculate the energy of the transition associated with each of the observed absorbances in Problem 2.2. Which requires the least energy for the associated transition?

2.4. Given that the concentrations of carvone and piperonal were 0.043 mmol dm^{-3} and 0.067 mmol dm^{-3} and that the path length for both spectra was 1 cm, calculate the molar absorptivity (ε) for each of the observed absorbances in Problem 2.2.

3
Infrared Spectroscopy

Aims

This chapter describes the uses of infrared (IR) spectroscopy in organic chemistry. By the end of the chapter you should:

- Be able to identify the functional groups present in a molecule from an analysis of its IR spectrum
- Be able to predict the region of the IR spectrum in which functional groups absorb
- Understand the relationship between stretching frequency, the reduced mass and the strength of a bond
- Understand the factors which influence the stretching frequency of a given vibration, *e.g.* hydrogen bonding and conjugation

3.1 Instrumentation

An IR spectrometer consists of an energy source (in this case a source of IR radiation, wavelength 2.5–15 μm), a sample holder, detector and plotter. In older instruments the IR beam was split into two equal intensity beams, one of which passed through the sample and the other did not. If the sample, or air, contains groups whose vibrational frequency corresponds to that of the IR radiation, it will absorb energy and an absorption peak will be seen. The beam which did not pass through the sample was used as the "background" or reference (absorptions due to atmospheric water, carbon dioxide, *etc*.), and the signal detected for this beam was subtracted from that of the sample, therefore making allowances for the atmospheric conditions.

Modern spectrometers employ a single beam, and the spectra are stored in a digital form (as in NMR: see Chapter 4). First, the background spectrum is obtained and stored, then the sample (plus background) spectrum is obtained. The instrument software then subtracts the background spectrum to give the sample spectrum. Modern instruments (again like NMR spectrometers) do not scan through the IR spectrum frequency range; the spectrum is obtained by irradiation with a single pulse of radiation, and all the frequencies are excited at once. The resulting spectrum is then Fourier transformed (a complex mathematical process which, thankfully, we do not need to know about to use IR instruments) to give the conventional looking spectrum. One advantage of this technique over the older instruments (which gave spectra that were the result of a single scan) is that several scans can be obtained and added together (in digital form). This addition of spectra results in an increase in the signal-to-noise ratio and spectra can be obtained on smaller amounts of material.

The method used for sample preparation depends upon the nature of the sample. Liquids are easily examined as films formed when one drop of the liquid is squeezed between two flat sodium chloride plates, which are transparent to IR radiation in the 4000–666 cm^{-1} region. Solids can be examined as solutions, mulls in Nujol, or as potassium bromide discs. For solutions, a 5% solution of the solid is introduced into a sodium chloride cell, which is usually 1 mm thick. The solvent employed should be reasonably transparent to IR, and the background should be obtained with the cell containing the solvent only.

Nujol mulls are prepared by finely powdering about 1 mg of the sample by grinding with a mortar and pestle then adding 1 drop of Nujol (a liquid hydrocarbon) and mixing thoroughly. The mull is then placed between sodium chloride plates as for liquid films. Problems with this method arise due to the insolubility of a wide range of organic compounds in Nujol and the presence of C–H absorption bands arising from the Nujol itself (Figure 3.1). Alternatively, about 1 mg of the sample is finely ground with about 100× its bulk of dry potassium bromide, then pressed into a thin KBr disc by means of a hydraulic press while connected to a vacuum pump to remove any water from the sample. This method, although slightly more time consuming than making a mull, gives spectra with no IR bands due to solvent as KBr, like NaCl, is again transparent in the region of interest.

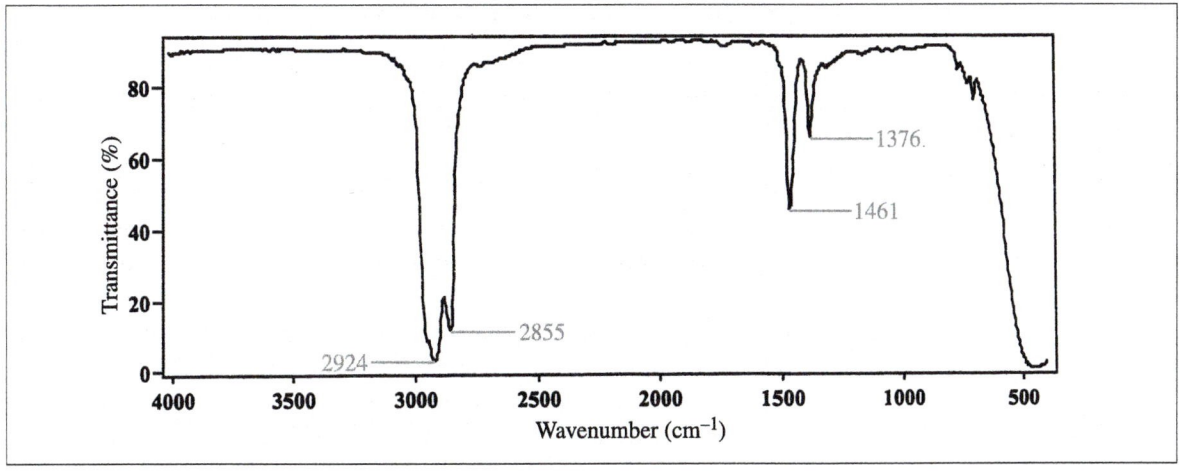

Figure 3.1 Nujol IR spectrum (liquid film, NaCl plates)

3.2 Selection Rules and Hooke's Law

IR spectroscopy corresponds to transitions between the vibrational energy levels of a molecule, involving the stretching or bending of bonds. For an organic chemist the region of interest for IR spectroscopy is of wavelength from 2.5 to 15 μm, which is normally expressed as the wavenumber range 4000–666 cm^{-1}, since this is the region in which the most diagnostically useful vibrations occur. Organic chemists usually begin their IR analysis starting from the peaks at the low-wavelength (high-wavenumber) end of the spectrum, *i.e.* from 4000 to 666 cm^{-1}.

Another type of vibrational spectroscopy, which can be used to study molecules without a changing dipole, is **Raman spectroscopy**, which also involves molecular vibrations and, in this case, an interaction between the molecular polarizability (the ease with which the electron cloud around a molecule can be distorted) and the IR radiation. A different selection rule applies – there must be a change in the polarizability during the vibration – so that some bands which are IR inactive are Raman active, *e.g.* the IR inactive C≡C stretch of acetylene is Raman active and appears at 2180 cm^{-1}. The use of Raman spectroscopy in organic chemistry is rare so it will not be discussed any further here.

IR transitions arise due to the interaction of the oscillating electric vector of the IR light with the oscillating dipole moment of the molecule (due to the molecular vibration). The selection rule for IR transitions – which must be obeyed if IR absorption is to occur for a particular vibration – states that the molecule must have a dipole moment and that there must be a change in the molecular dipole moment during the vibration, *i.e.* the dipole moment must be different at the extremes of the vibration. As an example of molecular dipole moments changing during vibrations, consider the centrosymmetric molecule acetylene (ethyne), H–C≡C–H. Acetylene has no dipole moment and the C≡C stretching vibration (a, Scheme 3.1) results in no change in the dipole moment; this vibration is, therefore, IR inactive (not observed in the IR spectrum). The symmetrical stretch of the two C–H bonds (b) also does not result in a change in the molecular dipole moment and so is also IR inactive. The asymmetric stretch (c) does, however, result in a change in the dipole moment and is IR active (the structures shown for each vibrational mode in Scheme 3.1 are the two extremes for each vibration).

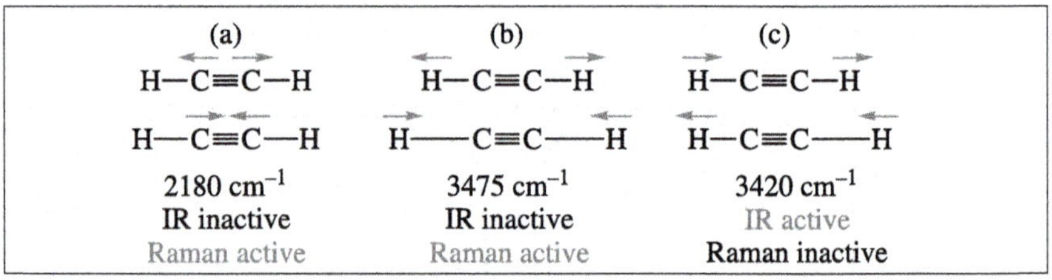

(a)	(b)	(c)
H—C≡C—H	H—C≡C—H	H—C≡C—H
H—C≡C—H	H——C≡C——H	H—C≡C——H
2180 cm^{-1}	3475 cm^{-1}	3420 cm^{-1}
IR inactive	IR inactive	IR active
Raman active	Raman active	Raman inactive

Scheme 3.1 Stretching vibrations of acetylene

As we shall see later, complex organic molecules have a large number of vibrational modes (including overtones, combinations and so-called "forbidden" bands), and these give rise to very complex spectra; however, as we shall also see, some vibrational modes can be attributed to individual functional groups (these are the most useful diagnostically to the organic chemist) and others to vibrations of the whole molecular structure. Those vibrational modes which can be attributed to individual functional groups (characteristic group vibrations) can be best described mathematically if we imagine the two bonded atoms as two vibrating masses connected by a spring (a simple harmonic oscillator). The frequency of the vibration for this system is then given by Hooke's law (equation 3.1):

$$v = \frac{1}{2\pi}\sqrt{\frac{k}{\mu}} \quad \text{or} \quad \bar{v} = \frac{1}{2\pi c}\sqrt{\frac{k}{\mu}} \tag{3.1}$$

where k is the force constant of the bond ($N\ m^{-1}$) – the stronger the bond the greater the value of k; $\bar{v}$ is the wavenumber (cm^{-1}); v is the frequency (Hz); c is the speed of light ($3 \times 10^{10}\ cm\ s^{-1}$); and μ is the reduced mass in kg (equation 3.2):

$N = kg\ m\ s^{-2}$.

$$\mu = \frac{m_1 \times m_2}{(m_1 + m_2)} \tag{3.2}$$

where $m_1 =$ relative atomic mass of $M_1 \times$ atomic mass unit, $m_2 =$ relative atomic mass of $M_2 \times$ atomic mass unit, and the atomic mass unit $= 1.66 \times 10^{-27}$ kg.

Looking at Hooke's law, we can see that for a strong bond (large k) connected to a light mass it predicts a high-frequency (or wavenumber) vibration, and this is in agreement with the observed wavenumbers for the O–H and N–H absorptions (3600–3200 cm^{-1}).

3.3   Characteristic Group Vibrations

As mentioned previously, complex organic molecules have a large number of vibrational modes, each of which will give rise to an IR absorption band as long as it involves a change in the molecular dipole moment. Some of these bands can only be attributed to vibrations of the whole molecular skeleton; these bands occur between 1500 and 1000 cm^{-1} and are characteristic of the molecule. This region is, therefore, known as the "fingerprint region" and, in the same way that our fingerprints are unique to us, the pattern of IR bands in this region is unique to a given molecule.

Some of the IR absorption bands can be attributed to the vibration of individual bonds, and these bands are the most diagnostically useful since they help in the identification of the functional groups present in the molecule. This is the most important use of IR spectroscopy for the organic chemist. In looking at these characteristic group vibrations, it is convenient to consider the IR spectrum as being divided into five regions:

4000–2300 cm^{-1} Vibrations of single bonds to H (C–H, O–H, N–H and S–H)

2300–1850 cm^{-1} Vibrations of triple and cumulative bonds

1850–1500 cm^{-1} Vibrations of C=X

1500–1000 cm^{-1} Fingerprint region

1000–666 cm^{-1} Unsaturated C–H bending

We will now consider the absorptions in each of these regions in detail, but an indication of where the common functional groups absorb is given in Box 3.1.

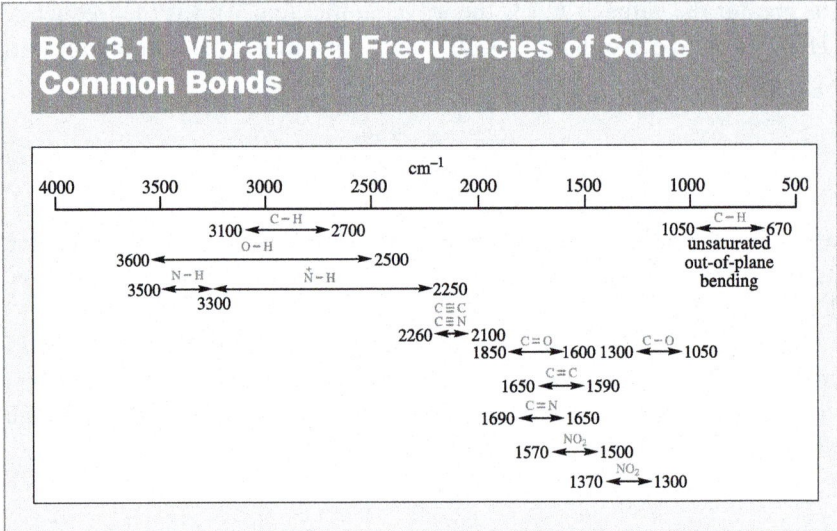

Box 3.1   Vibrational Frequencies of Some Common Bonds

3.3.1   4000–2300 cm^{-1}: Vibrations of Single Bonds to H

Absorption bands in this region correspond to the stretching vibrations of single bonds to hydrogen: C–H, O–H, N–H and S–H.

C–H (3100–2700 cm^{-1})

C–H vibrations are rarely diagnostically useful, since all organic molecules contain C–H bonds and there is little difference in the stretching frequencies of the C–H bonds in most environments. In addition, C–H stretches are usually weak due to the small dipole moment of this bond. We can often, however, distinguish between a C–H in an unsaturated group (3100–3000 cm^{-1}) and a C–H in a saturated group (about 2900 cm^{-1}), but the most useful C–H stretches are those for an aldehyde and an alkyne (Scheme 3.2). For example, in the spectrum of benzaldehyde (Figure 3.2) we can see that there are two moderately strong C–H stretches at 2820 and 2740 cm^{-1}.

Scheme 3.2 Diagnostically useful C–H stretches

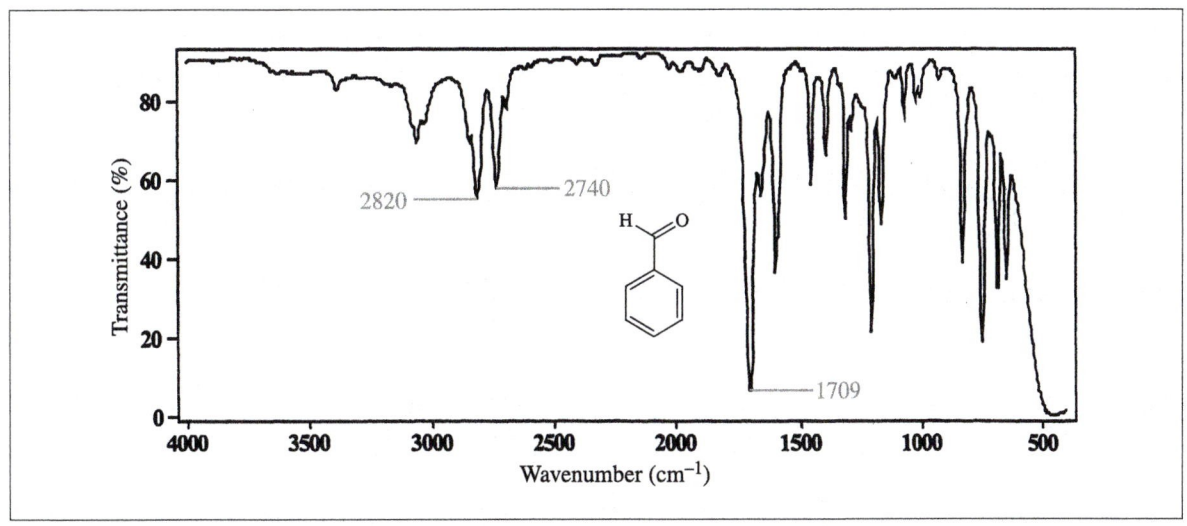

Figure 3.2   IR spectrum of benzaldehyde (liquid film, NaCl plates)

In general, strong absorptions above 3100 cm^{-1} indicate either O–H or N–H vibrations.

O–H (3600–2500 cm^{-1})

O–H vibrations are generally strong and are very useful in identifying functional groups containing the hydroxyl group, but we should always be aware that the presence of water in a sample will also give a strong O–H absorption.

This functional group is present in phenols (ArOH), alcohols (ROH) and carboxylic acids (RC(O)OH). Unfortunately, the wavenumber of the absorption is not a reliable indicator of the functional group present, since the position of the O–H absorption is very sensitive to hydrogen bonding. In general, the more concentrated a solution, the greater the extent of hydrogen bonding and, since hydrogen bonding weakens the O–H bond, the smaller the value of k (the force constant) and so the lower the frequency (or wavenumber) of the vibration.

Examples of the type of peaks which arise due to absorption by the O–H group can be seen in the IR spectrum of 4′-hydroxyacetophenone (Figure 3.3), in which there are broad O–H peaks centered on 3304 cm^{-1} (non-hydrogen bonded) and 3158 cm^{-1} (due to hydrogen bonding), and paracetamol (Figure 3.4), in which there is an O–H absorption at 3161 cm^{-1}.

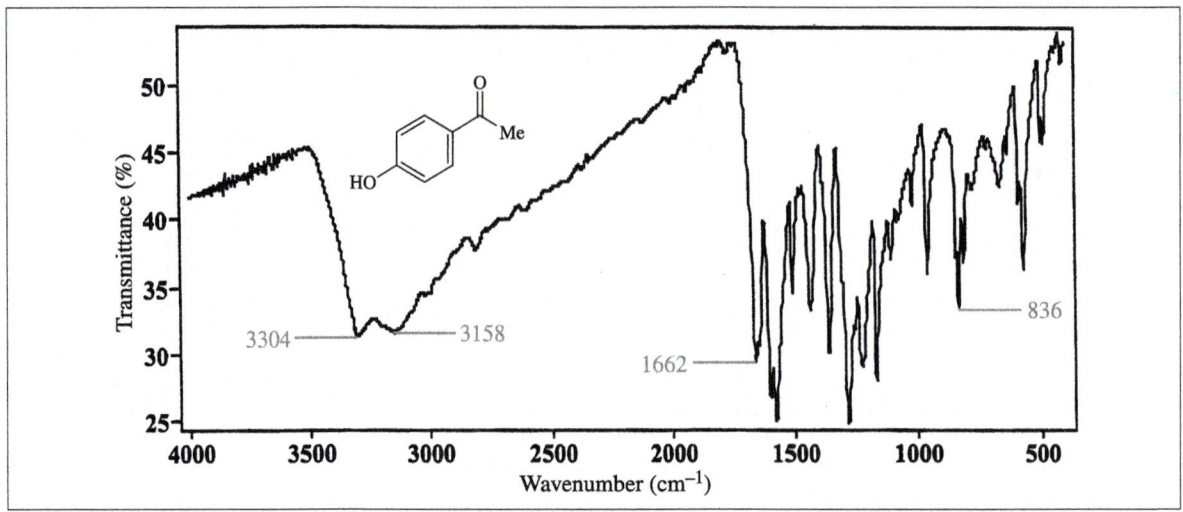

Figure 3.3   IR spectrum of 4′-hydroxyacetophenone (KBr disc)

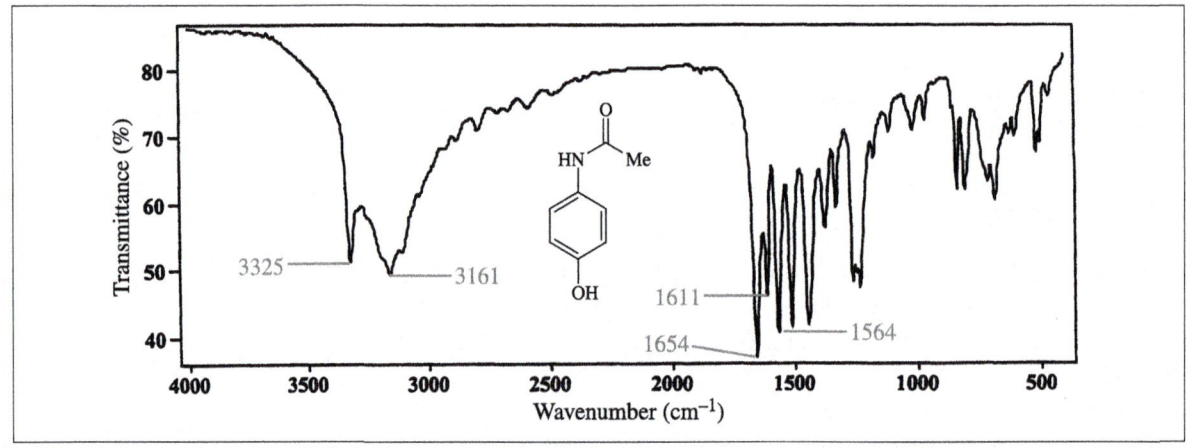

Figure 3.4   IR spectrum of paracetamol (KBr disc)

Scheme 3.3   Hydrogen-bonded dimer of a carboxylic acid

Carboxylic acids (RCO_2H) often form dimers (Scheme 3.3) and tend to have a higher degree of hydrogen bonding, so they usually absorb towards the lower end of this region. Because of this hydrogen bonding, carboxylic acids tend to have a number of weak absorptions (corresponding to different bond strengths) which cover a range of frequencies such that they normally give rise to broad absorptions – often with some fine structure just evident – between 3200 and 2500 cm^{-1}. Examples of the resultant broad O–H absorption peaks can be seen in the IR spectra of benzoic acid (Figure 3.5), in which there are multiple peaks between 3010 and 2560 cm^{-1}, and aspirin (Figure 3.6), in which there is a similar pattern between 3005 and 2550 cm^{-1}.

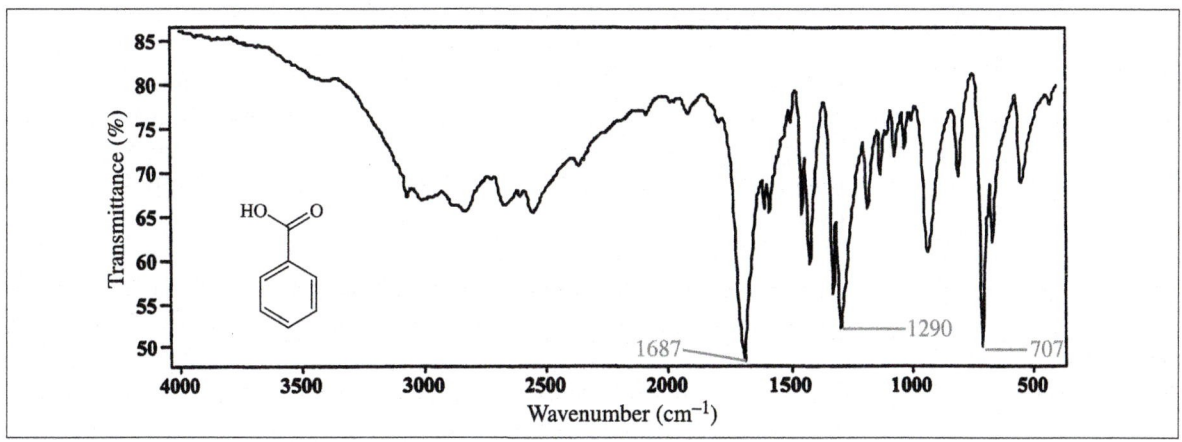

Figure 3.5 IR spectrum of benzoic acid (KBr disc)

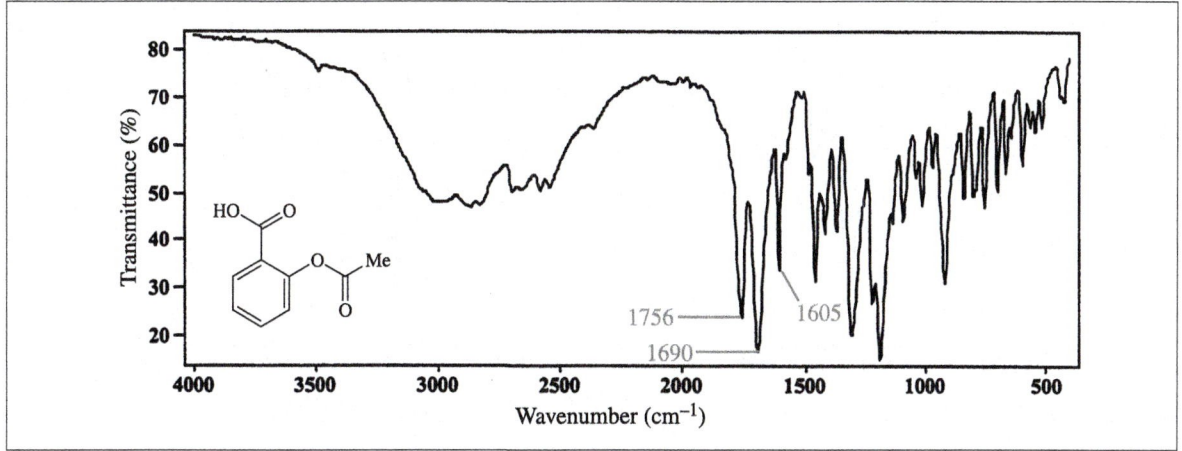

Figure 3.6 IR spectrum of aspirin (KBr disc)

Solids or dilute solutions of phenols, alcohols and even carboxylic acids have a lesser degree of hydrogen bonding and so absorb in the $3600-3200$ cm^{-1} region. If the presence of a carboxylic acid is suspected, then there will, of course, be an absorption for the associated carbonyl group at $1730-1680$ cm^{-1}.

N–H ($3500-3300$ cm^{-1})

N–H absorptions appear in much the same region as O–H, but it is usually possible to distinguish the N–H absorptions as they are generally sharper owing to a smaller degree of hydrogen bonding. Once again, these absorptions are reasonably strong, and we can easily distinguish primary amines (RNH$_2$) and primary amides (RCONH$_2$) from secondary amines (R$_2$NH) and secondary amides (RCONHR). Primary amines and amides

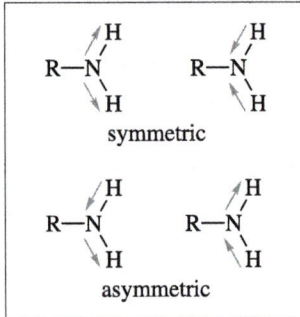

Scheme 3.4 Symmetric and asymmetric stretches of a primary amine

have two (very often three) stretches in this region corresponding to the symmetrical and asymmetrical stretches (Scheme 3.4) of the two N–H bonds in these functional groups, while secondary amines and amides have only one N–H stretching vibration. In distinguishing N–H from O–H vibrations, it is useful to note that for an N–H stretching vibration there should be a corresponding bending vibration at 1650–1550 cm^{-1} (normally weak).

The IR spectrum of aniline (Figure 3.7) shows a typical N–H absorption pattern for a primary amine, with stretches at 3432, 3355 and 3214 cm^{-1}, while that of paracetamol (Figure 3.4) shows the single stretch observed for a secondary amide at 3325 cm^{-1} and we can also see the contrast between this sharp peak and the much broader O–H at 3161 cm^{-1}. The spectrum of benzocaine (Figure 3.8) also shows the three bands at 3423, 3342 and 3221 cm^{-1} which characterize a primary amine.

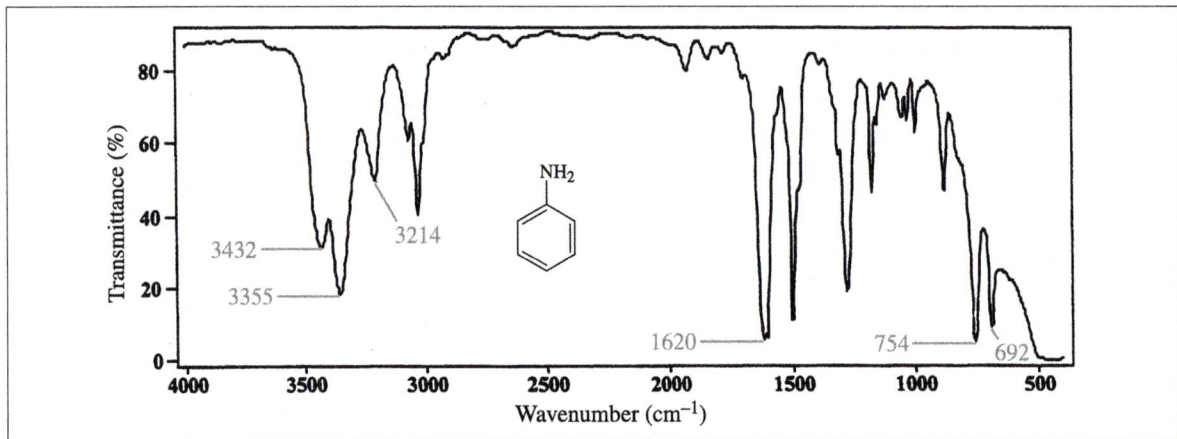

Figure 3.7   IR spectrum of aniline (liquid film, NaCl plates)

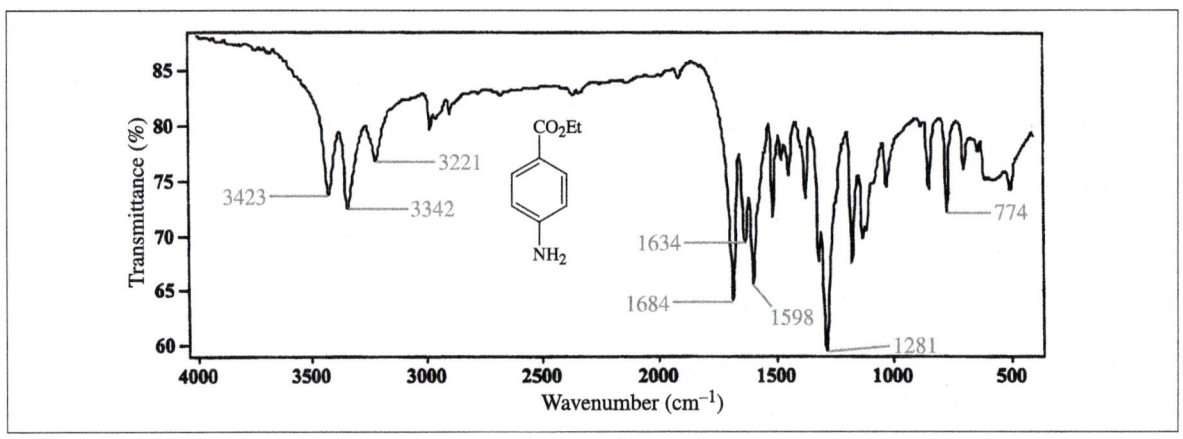

Figure 3.8   IR spectrum of benzocaine (KBr disc)

N^+–H (3300–2250 cm^{-1})

For salts of amines we also obtain strong bands for both the stretching vibrations and the associated bending modes (Table 3.1).

Table 3.1 Stretching and bending vibrations of ammonium salts

Ammonium salt	Stretching vibration/cm^{-1}	Bending vibration/cm^{-1}
RNH_3^+	~3000	1600–1500
$R_2NH_2^+$	2700–2250	1600–1550
R_3NH^+	2700–2250	not diagnostically useful

The spectrum of the zwitterionic amino acid L-tyrosine shows the effect of protonation of the amino group, with the N^+–H stretching vibration occurring over a broad range (3100–2600 cm^{-1}), while the carboxylate group stretching vibration is weakened and comes at around 1589 cm^{-1}, along with the N^+–H bending and C=C of the aromatic ring (Figure 3.9).

To what functional group does the band at 3199 cm^{-1} in Figure 3.9 correspond?

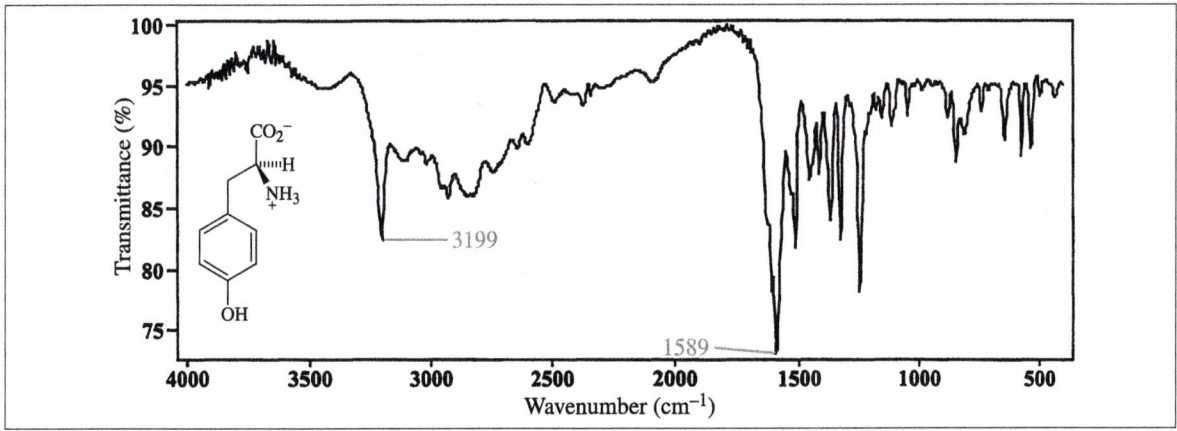

Figure 3.9 IR spectrum of L-tyrosine (KBr disc)

S–H (2700–2400 cm^{-1})

The S–H absorption is typically weak and occurs at a lower wavenumber (frequency) than the corresponding O–H vibration because of the higher atomic mass of sulfur (and so increased reduced mass of SH). The difference in reduced mass, however, does not completely account for the reduction in the stretching frequency of the S–H bond as this bond is also generally weaker (smaller k) than the corresponding O–H bond (typical bond dissociation energies: OH, 460 kJ mol^{-1}; SH, 340 kJ mol^{-1}).

$$\bar{v} = \frac{1}{2\pi c}\sqrt{\frac{k}{\mu}}$$

3.3.2 2300–1850 cm^{-1}: Vibrations of Triple and Cumulative Bonds

Absorptions in this region typically arise due to vibrations of functional groups containing triple or cumulative bonds.

When discussing the IR selection rules, we saw that the C≡C bond of acetylene absorbs at 2180 cm^{-1} (although this particular absorption band is IR inactive). Table 3.2 gives the stretching vibrations for alkynes and other triple and cumulative bonds which absorb in this region.

Table 3.2   Stretching vibrations of triply and cumulatively bonded functional groups

Functional group	Stretching vibration/cm^{-1}	
–C≡C– (alkynes)	Terminal R–C≡C–H	2140–2100
	Internal R–C≡C–R	2260–2190
–C≡N (nitriles)	2260–2100	
$-\overset{+}{N}\equiv N$ (diazonium salts)	2280–2240	
$-N=\overset{+}{N}=\overset{-}{N}$ (azides)	2160–2100	
$-\overset{\mid}{C}=\overset{+}{N}=\overset{-}{N}$ (diazo compounds)	2050–2000	
–N=C=N– (carbodiimides)	2150–2110	
$-\overset{\mid}{C}=C=\overset{\mid}{C}-$ (allenes)	~1950	
$\overset{\mid}{C}=C=O$ (ketenes)	~2150	
CO_2	2350	

Most of these absorption bands are moderate to weak; in particular, the alkyne bands decrease in intensity with increasing symmetry in the molecule. An example of a strong nitrile absorption can be seen in the spectrum of benzonitrile (Figure 3.10) at 2228 cm^{-1}.

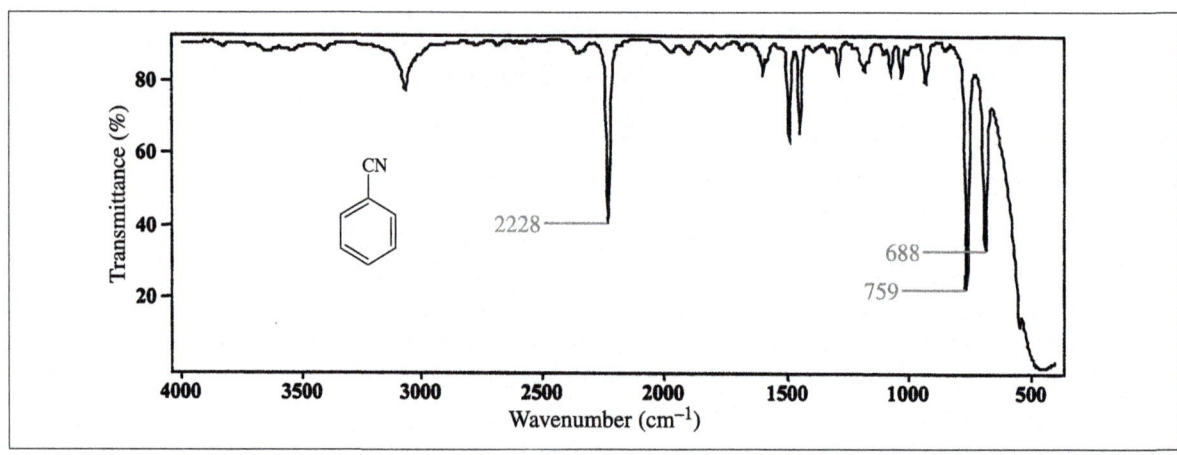

Figure 3.10   IR spectrum of benzonitrile (liquid film, NaCl plates)

3.3.3   1850–1500 cm^{-1}: Vibrations of C=X

This is probably the most useful diagnostic region as it includes the stretching vibrations for those common functional groups which contain a double bond to carbon (C=C, C=N and C=O). The absorptions for carbonyl groups are particularly useful, since they are strong and their position is very sensitive to the type of functional group present, as well as to any hydrogen bonding or conjugation present in the molecule.

Box 3.2 gives an indication of the typical ranges for the stretching vibrations of carbonyl-containing functional groups. In general, the more electronegative the group X in RCOX, the stronger the C=O bond, and so the higher the wavenumber (frequency) for the stretching vibration. We can partly explain this increase in wavenumber by imagining a resonance form for RCOX in which the C–X bond is broken as a result of the electron-demand of the X group, as shown in Scheme 3.5.

Scheme 3.5 Resonance showing strong carbonyl group absorption when X is strongly electronegative

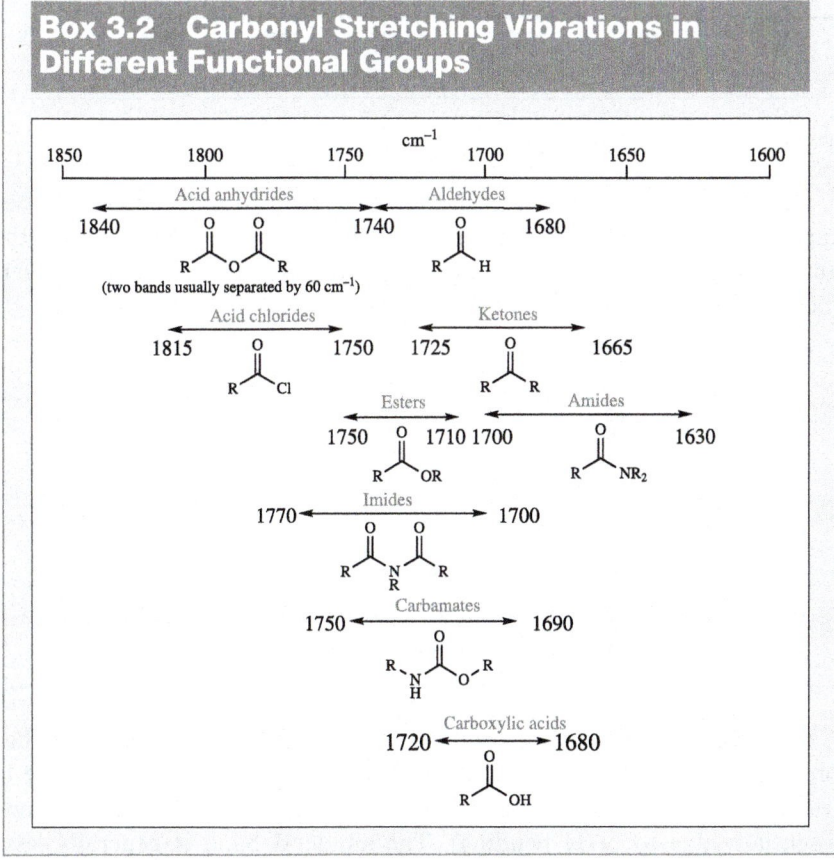

Box 3.2 Carbonyl Stretching Vibrations in Different Functional Groups

Obviously, the more electronegative the group X, the more likely this is to happen and, since one of the resonance forms has a C≡O (and hence a larger value for the force constant, k), this will lead to a stronger C=O bond, and so increased stretching frequency. For example, the two carbonyl absorptions in acetic anhydride (ethanoic anhydride; in which the X group is the very electronegative $CH_3CO_2^-$) can be seen at 1822 and 1762 cm^{-1} in Figure 3.11. This separation by 60 cm^{-1} is very common for these bands, because of the asymmetric and symmetric C=O stretches.

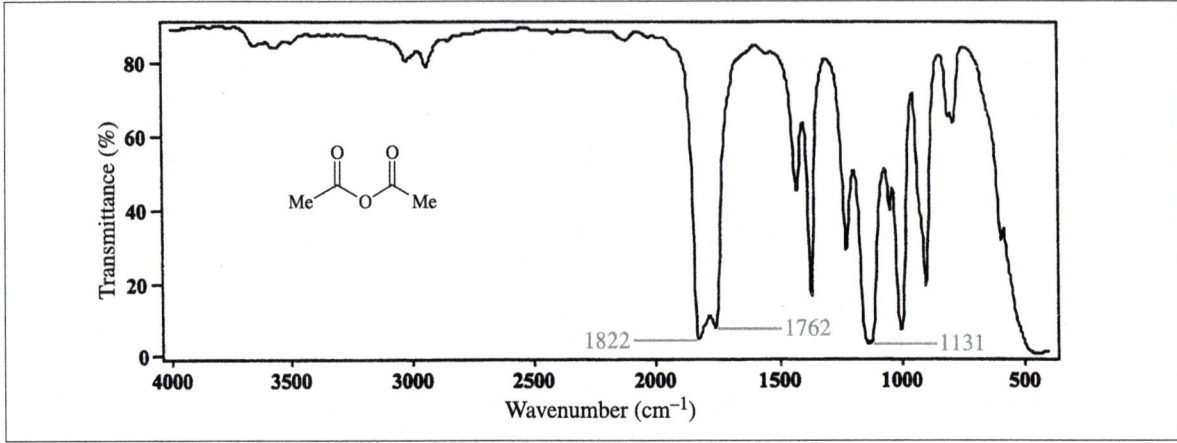

Figure 3.11   IR spectrum of acetic anhydride (liquid film, NaCl plates)

On the other hand, electron-donating groups (particularly those capable of resonance stabilization) will decrease the C=O bond strength and so lead to a decrease in the stretching frequency (wavenumber) (Scheme 3.6).

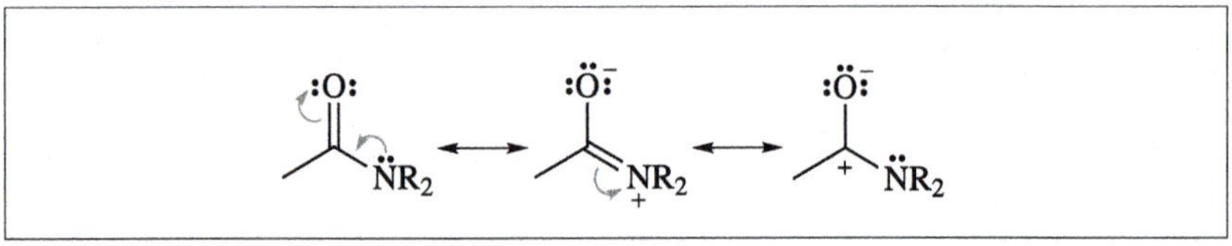

Scheme 3.6   Resonance forms of an amide. The C–O bond has partial single-bond character, so a smaller k

In the spectrum of paracetamol (Figure 3.4), identify the C=O absorption for the secondary amide group.

One interesting feature of primary and secondary amide spectra is the presence of two absorption bands in this region – referred to as amide I (essentially the carbonyl band) and amide II (a combination band essentially due to N–H bending). The amide II band is generally less intense than amide I and is often about 80 cm^{-1} lower than the amide I band.

Conjugation and hydrogen bonding also serve to weaken the C=O bond, through a decrease in the bond order, and so lead to a decrease in the stretching frequency. As we can see in Scheme 3.7, there is a resonance form for a conjugated carbonyl group in which the C–O bond is a single bond. Overall, then, the conjugated carbonyl group has a lower bond order than an isolated carbonyl group, so it has a smaller k and hence a smaller stretching frequency. Conjugation does, however, increase the dipolar nature of the bond, and so usually leads to an increase in the intensity of the carbonyl absorption. The effect of conjugation on a carbonyl absorption can be seen in the IR spectra of the esters ethyl acetate (ethyl ethanoate, Figure 3.12) and ethyl benzoate (Figure 3.13).

Identify the C=O absorptions in the spectra of these esters and note the effect of conjugation on the C=O stretch. You should also identify, and compare the wavenumbers of, the C=O absorptions of 2-methylpropanal (Figure 3.14) and the conjugated aldehyde benzaldehyde (Figure 3.2).

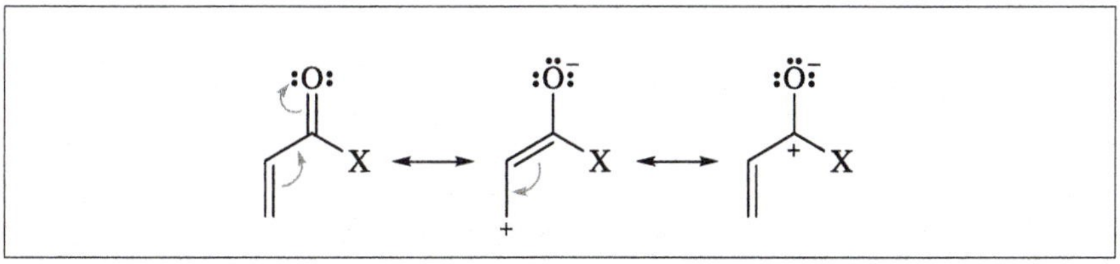

Scheme 3.7 Resonance forms of a conjugated carbonyl group. The C–O bond has partial single-bond character, so a smaller k

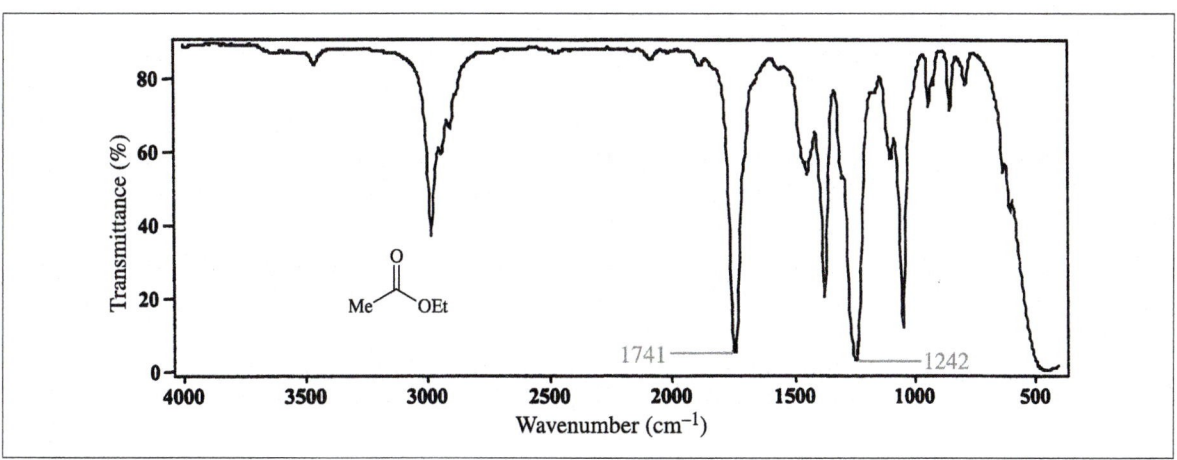

Figure 3.12 IR spectrum of ethyl acetate (liquid film, NaCl plates)

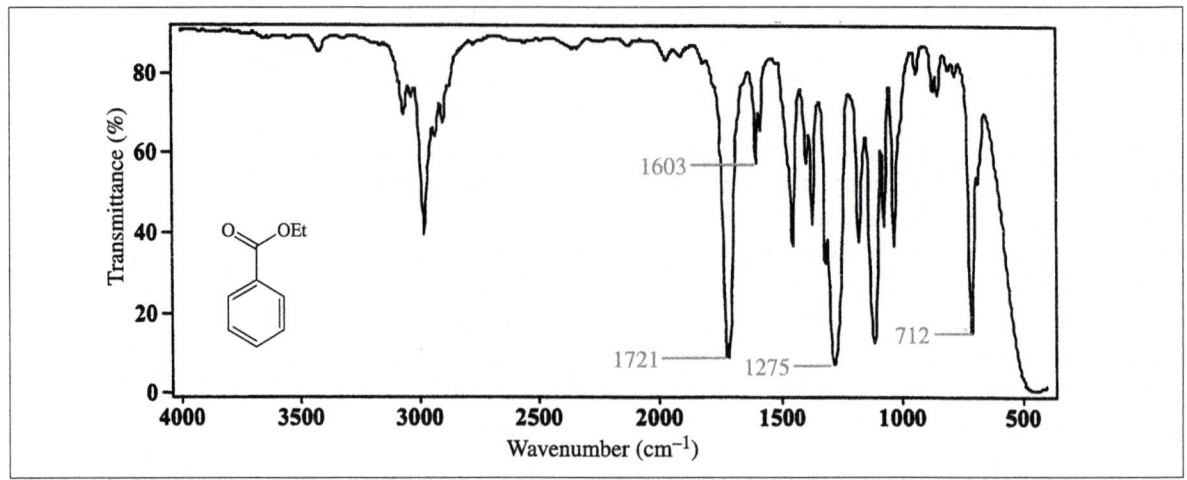

Figure 3.13 IR spectrum of ethyl benzoate (liquid film, NaCl plates)

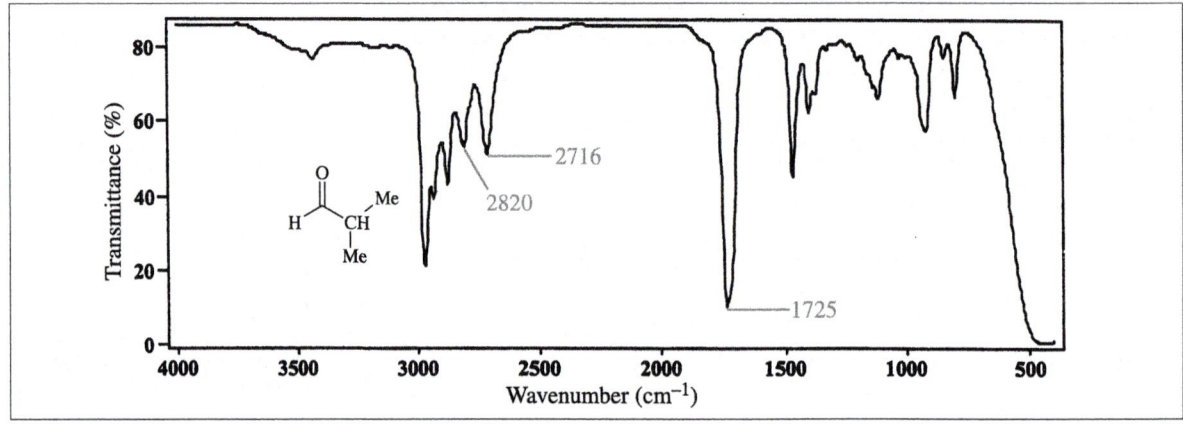

Figure 3.14 IR spectrum of 2-methylpropanal (liquid film, NaCl plates)

Hydrogen bonding also leads to a decrease in the carbonyl stretching frequency, since we can think of the hydrogen bond as utilizing some of the carbonyl bond electron density. This results in a lower bond order for this bond, and a consequently lower stretching frequency. One interesting aspect of the effect of hydrogen bonding on the IR spectrum of carbonyl compounds is the effect of intra- and intermolecular hydrogen bonding. Intramolecular hydrogen bonding (within one molecule) is *independent of the concentration* of the sample and so the position of the carbonyl band will not vary with concentration. Intermolecular hydrogen bonding, on the other hand, is *dependent upon the concentration*, since the more concentrated the sample, the greater the extent of the hydrogen bonding, and so the position of the carbonyl band will vary with concentration. The more concentrated the sample, the lower the stretching frequency.

The effect of increasing concentration (and so increasing extent of hydrogen bonding) on the solution spectrum of 4′-hydroxyacetophenone can be seen in Figure 3.15 and Scheme 3.8b, in which the C=O absorption shifts to lower frequency (wavenumber) upon increasing the concentration of the solution. The C=O absorption in 4′-hydroxyaceto-phenone initially occurs at 1672 cm^{-1} (Figure 3.15a), but is shifted to 1665 cm^{-1} upon increasing the concentration (Figure 3.15b). The C=O absorption in 2′-hydroxyacetophenone (Figure 3.16a and 3.16b, and Scheme 3.8a) appears at lower wavenumber than that for 4′-hydroxy-acetophenone at approximately equivalent concentrations (compare Figures 3.15b and 3.16a) and, as expected for intramolecular hydrogen bonding, is independent of concentration (1641 cm^{-1}). Also evident from these spectra is the fact that the increase in hydrogen bonding in Figure 3.15b has shifted the broad O–H absorption to lower wavenumber (from 3330 to 3320 cm^{-1}), but the absorption corresponding to non-hydrogen-bonded O–H at 3584 cm^{-1} is unaffected.

Q Other than the shift to lower wavenumber of the C=O and O–H bands, what other feature of Figure 3.15b indicates increased hydrogen bonding?

A The increase in the intensity of the broad O–H absorption (at 3330 cm^{-1} in Figure 3.15a and 3320 cm^{-1} in Figure 3.15b).

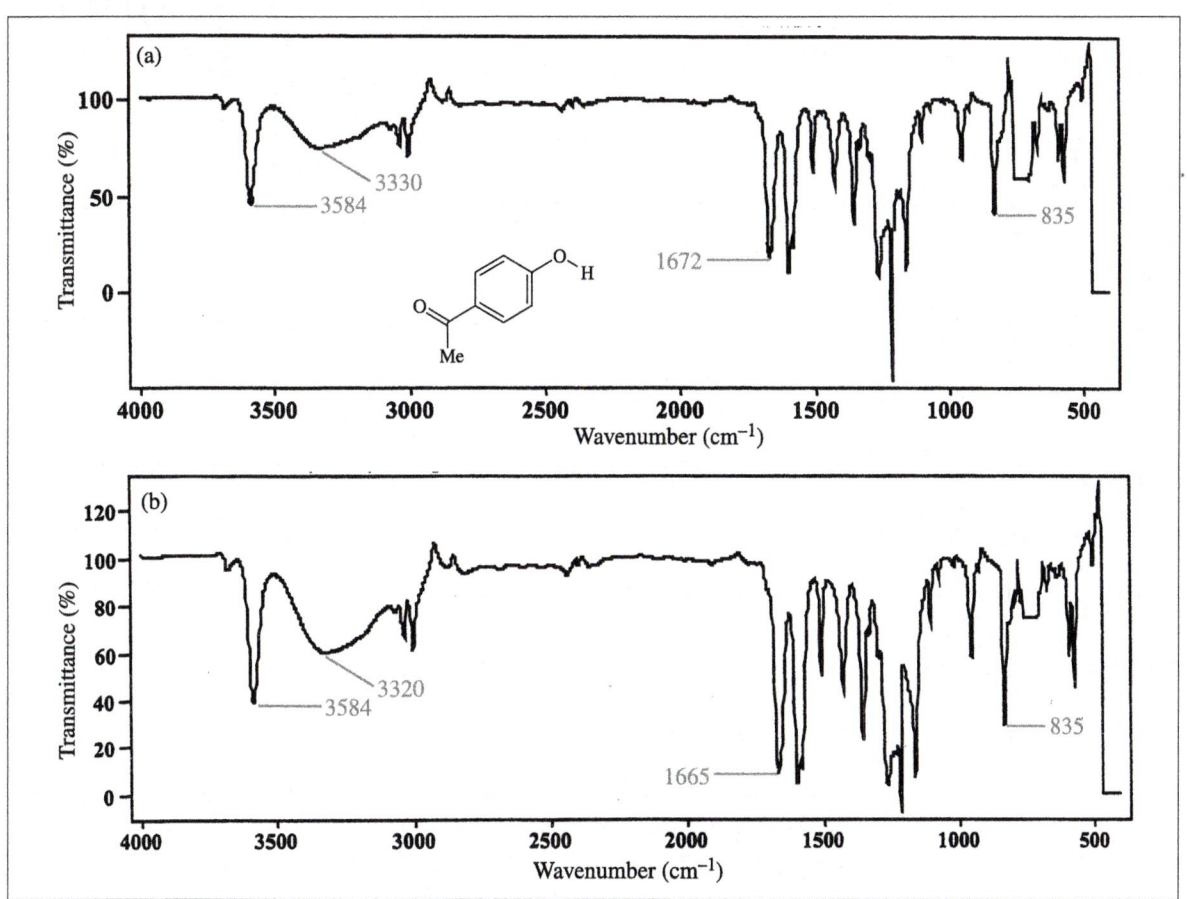

Figure 3.15 IR spectra of 4′-hydroxyacetophenone: (a) 7 mg cm^{-3} solution in chloroform (trichloromethane); (b) 14 mg cm^{-3} solution in chloroform

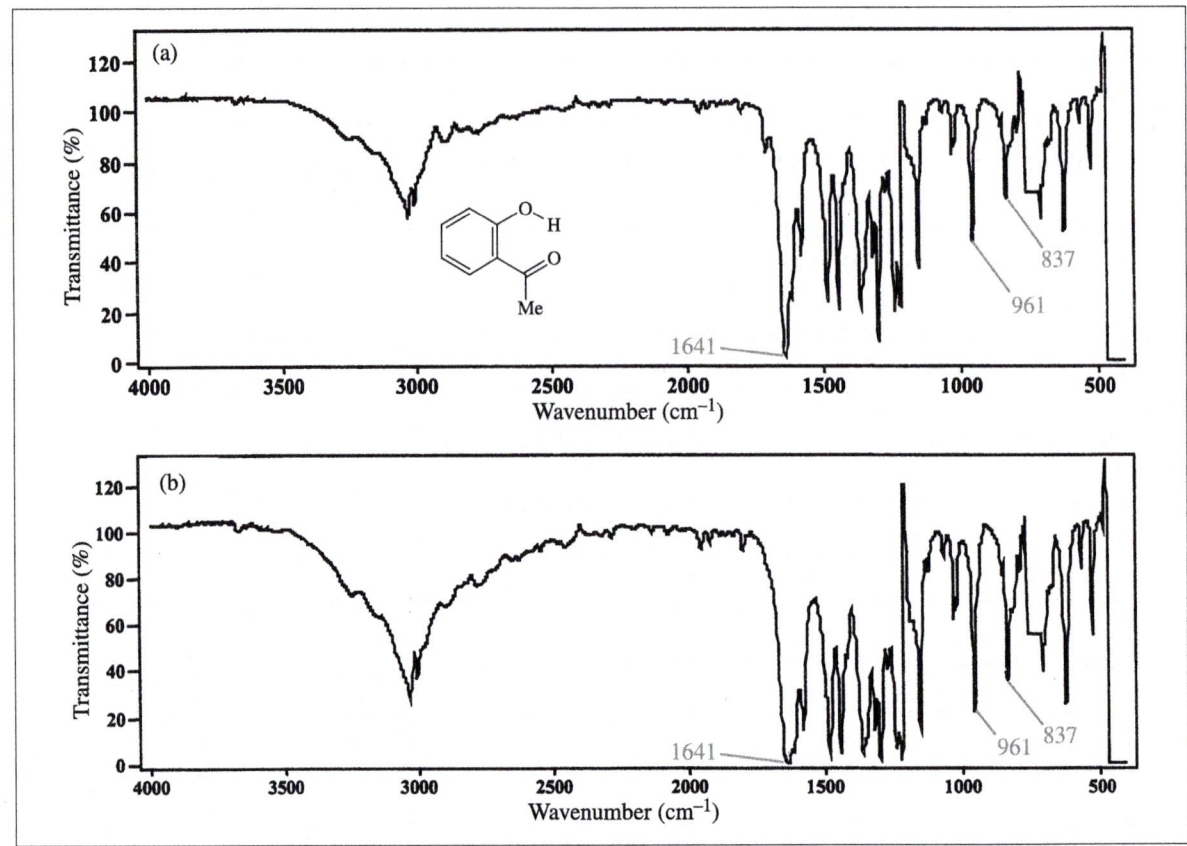

Figure 3.16 IR spectra of 2′-hydroxyacetophenone: (a) 12 mg cm^{-3} solution in chloroform (trichloromethane); (b) 24 mg cm^{-3} solution in chloroform

Scheme 3.8 Inter- and intra-molecular hydrogen bonding: (a) 2′-hydroxyacetophenone showing intramolecular hydrogen bonding (not concentration dependent); (b) 4′-hydroxy-acetophenone showing inter-molecular hydrogen bonding (concentration dependent)

The effect of ring strain on the stretching frequency of vibrations, particularly those of carbonyl groups, can be seen in the stretching frequencies of lactones (cyclic esters) and lactams (cyclic amides), as shown in Box 3.3. As can be observed, the smaller the ring becomes, the more strained the ring (as the bond angles become increasingly removed from the idealized angle of 120° for an sp^2 hybridized carbon atom and the amount of eclipsing interactions with neighbouring groups increases), and the higher the stretching frequency of the carbonyl group in both the lactones and lactams. The value for the six-membered rings is usually close to that for the acyclic systems, since this ring (and larger rings) have very little ring strain.

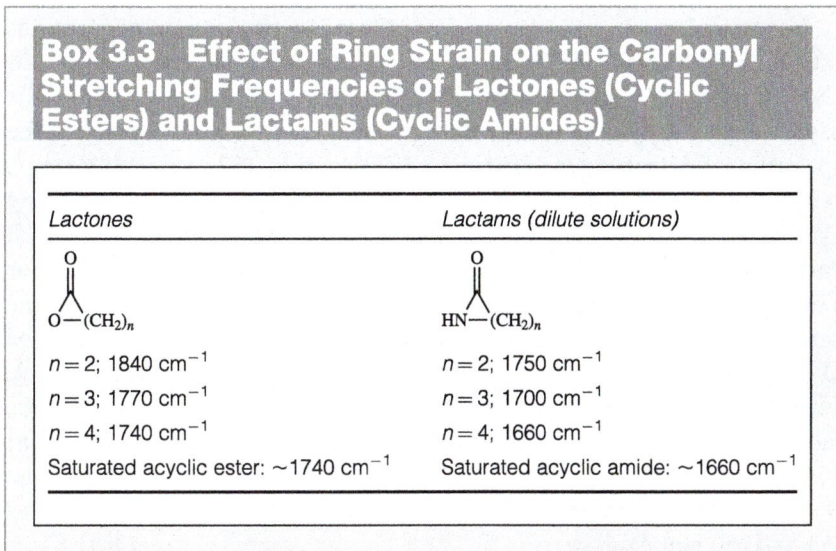

Box 3.3   Effect of Ring Strain on the Carbonyl Stretching Frequencies of Lactones (Cyclic Esters) and Lactams (Cyclic Amides)

Lactones	Lactams (dilute solutions)
$n = 2$; 1840 cm^{-1}	$n = 2$; 1750 cm^{-1}
$n = 3$; 1770 cm^{-1}	$n = 3$; 1700 cm^{-1}
$n = 4$; 1740 cm^{-1}	$n = 4$; 1660 cm^{-1}
Saturated acyclic ester: ~1740 cm^{-1}	Saturated acyclic amide: ~1660 cm^{-1}

Also present in this region are the C=C and C=N stretching vibrations (Scheme 3.9), but these are generally much weaker than the carbonyl absorptions.

Examples of the C=C stretches in aromatic rings can be seen in the spectra of paracetamol (Figure 3.4) at 1611 cm^{-1}, aspirin (Figure 3.6) at 1605 cm^{-1} and benzocaine (Figure 3.8) at 1634 cm^{-1}. The C=N group of oximes (C=NOH) absorbs between 1680 and 1650 cm^{-1}, and the corresponding O–H absorption occurs between 3600 and 3570 cm^{-1} (often lowered by hydrogen bonding).

Finally, in this region are the N–H bending vibrations (1640–1550 cm^{-1}), which will have corresponding stretching vibrations between 3600 and 3200 cm^{-1}. These absorptions are often difficult to distinguish from C=C vibrations. The N–H vibration of the secondary amide paracetamol is probably responsible for the absorption at 1564 cm^{-1} (Figure 3.4), while that of aniline (Figure 3.7) may be partly responsible (in addition to the C=C stretch) for the peak at 1620 cm^{-1}.

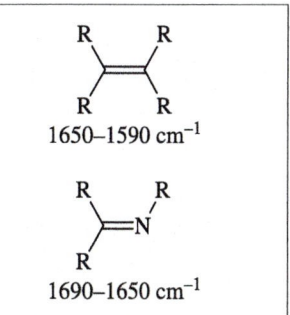

Scheme 3.9 C=C and C=N vibration stretching frequencies

C=C stretching vibrations will be absent if the molecule is symmetrical.

3.3.4 1500–1000 cm⁻¹: Fingerprint Region

Absorptions in this region are largely due to the vibrations of the whole molecular skeleton, and this region is known as the "fingerprint region". Because every organic molecule has absorptions in this region, it is often difficult to pick out any functional group vibrations that happen to occur in this region, but there are a few which are sufficiently strong to stand out from the crowd of those belonging to the molecular skeleton.

In particular, nitro compounds usually have two strong absorptions (at 1570–1500 and 1370–1300 cm⁻¹), whereas nitroso compounds (N=O) have only one absorption (1600–1450 cm⁻¹). For example, the asymmetric and symmetric stretching vibrations of the nitro group in the spectrum of nitrobenzene (Figure 3.17) occur at 1525 and 1348 cm⁻¹. Another strong absorption is the C–O absorption of ethers (including epoxides), alcohols, esters and carboxylic acids (1300–1050 cm⁻¹), which can be seen at 1281 cm⁻¹ in the spectrum of benzocaine (Figure 3.8) or 1131 cm⁻¹ in the spectrum of acetic anhydride (Figure 3.11). For the same reason that the S–H bond absorbs at lower frequency than the O–H bond, thioketones (C=S) absorb in this region (1200–1050 cm⁻¹). Finally in this region are the S=O stretches of a variety of compounds, including sulfoxides, sulfones and sulfonic acids, which occur at 1350–1010 cm⁻¹, *e.g.* dimethyl sulfoxide (Figure 3.18) in which the S=O stretch appears at 1057 cm⁻¹, and the P=O stretch of phosphates (1300–1250 cm⁻¹).

> The two absorptions of nitro groups are usually about 1530 and 1350 cm⁻¹.

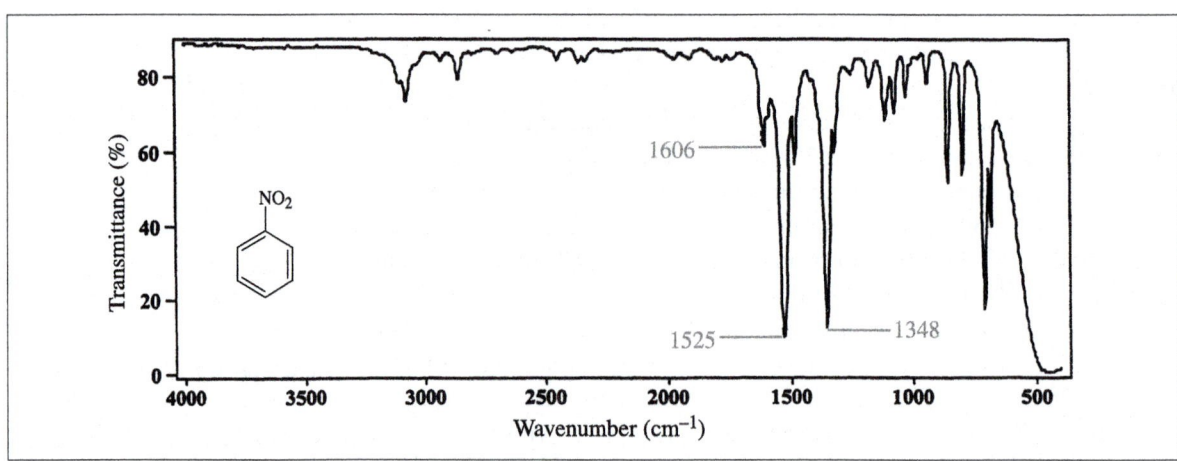

Figure 3.17 IR spectrum of nitrobenzene (liquid film, NaCl plates)

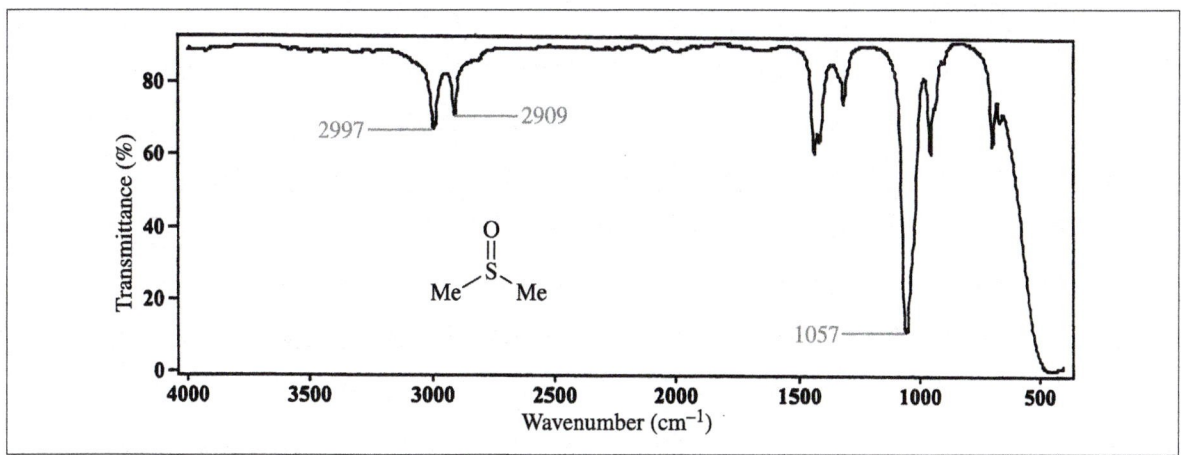

Figure 3.18 IR spectrum of dimethyl sulfoxide (liquid film, NaCl plates)

3.3.5 1000–666 cm^{-1}: Unsaturated C–H Bending

All of the vibrations in this region can be attributed to the out-of-plane bending vibrations of the C–H bonds of unsaturated groups, particularly aromatic rings (see Table 3.3). Many experienced organic chemists can use these vibrations to distinguish between the substitution patterns of di- and trisubstituted benzene rings, but this is not something which is recommended for beginners. A much easier and foolproof method of distinguishing these substitution patterns is to concentrate upon the aromatic region of the ^{1}H NMR spectrum. Monosubstituted aromatic rings are the most readily distinguished, since they usually have two strong absorptions in this region (typically at 740 and 690 cm^{-1}), whereas a 1,2-disubstituted (*ortho*) ring has only the one band (see Box 3.4).

Table 3.3 Out-of-plane bending vibrations of the C–H bonds in unsaturated groups

Unsaturated group	Bending vibration/cm^{-1}
Alkenes	995–710
cis-Alkenes	710–670
trans-Alkenes	970–960
Aromatics	900–690

Box 3.4 Out-of-plane C–H Bending Vibration Frequencies of Mono- and Disubstituted Aromatic Rings

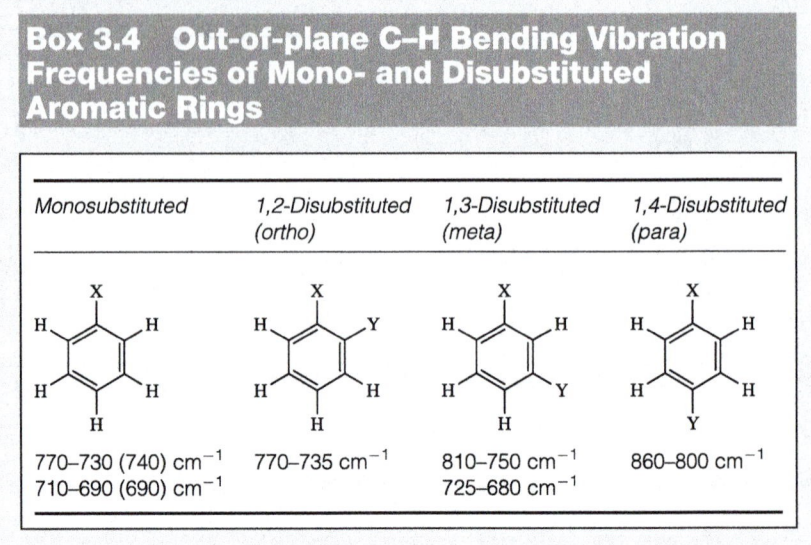

Monosubstituted	1,2-Disubstituted (ortho)	1,3-Disubstituted (meta)	1,4-Disubstituted (para)
770–730 (740) cm^{-1} 710–690 (690) cm^{-1}	770–735 cm^{-1}	810–750 cm^{-1} 725–680 cm^{-1}	860–800 cm^{-1}

The IR spectrum of the monosubstituted benzonitrile (Figure 3.10) gives two C–H out-of-plane bending peaks (at 759 and 688 cm^{-1}), that of aniline (Figure 3.7) gives peaks at 754 and 692 cm^{-1}, but ethyl benzoate (Figure 3.13) and benzoic acid (Figure 3.5) give only one strong C–H out-of-plane bending peak (at 712 and 707 cm^{-1} respectively). Finally, the 1,4-disubstituted (*para*) 4′-hydroxyacetophenone (Figure 3.3) gives one C–H out-of-plane bending peak at 836 cm^{-1}, the 1,3-disubstituted (*meta*) 3′-hydroxyacetophenone (Figure 3.19) gives two peaks (at 795 and 683 cm^{-1}) and the 1,2-disubstituted (*ortho*) 2′-hydroxyacetophenone (Figure 3.20) has the main peak at 755 cm^{-1}.

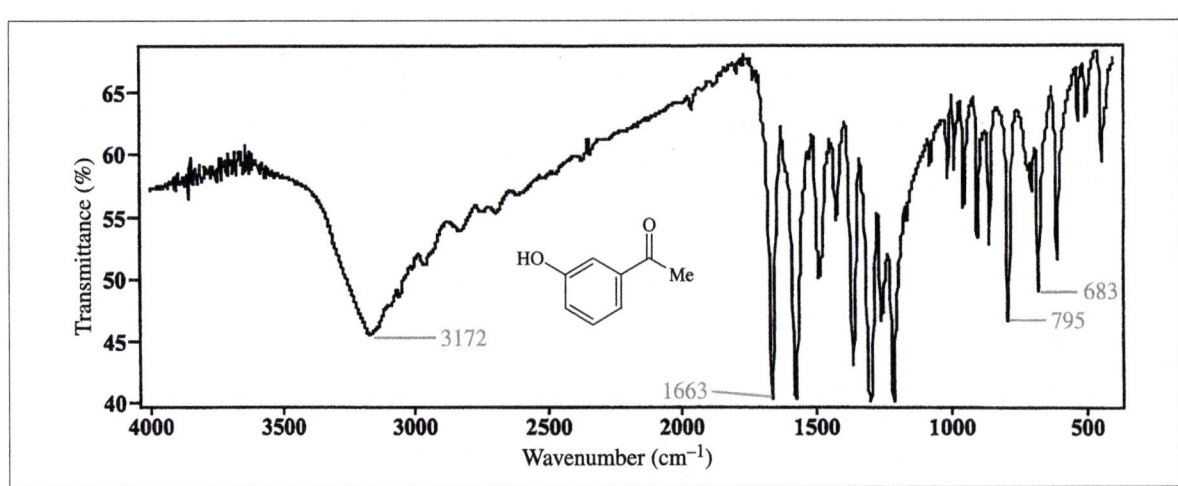

Figure 3.19 IR spectrum of 3′-hydroxyacetophenone (KBr disc)

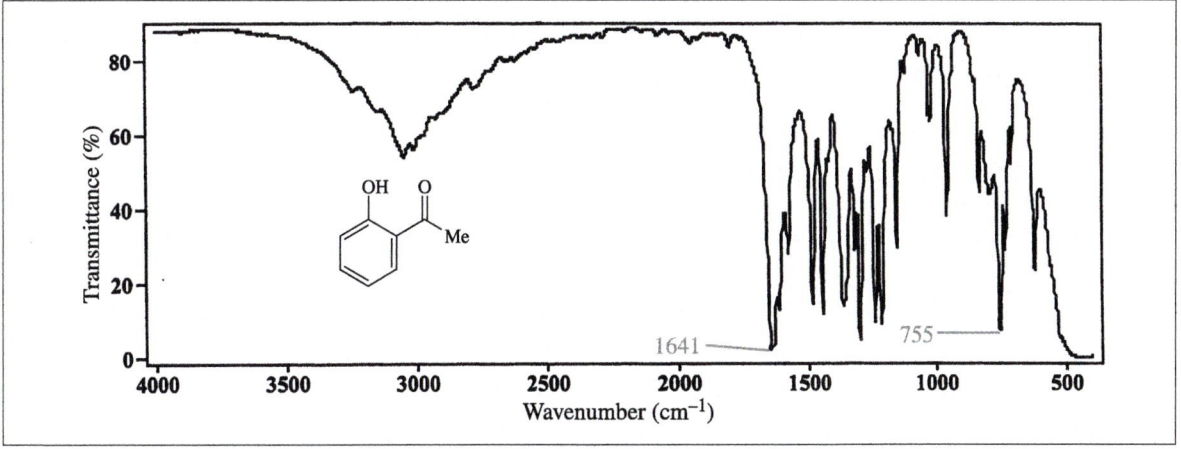

Figure 3.20   IR spectrum of 2′-hydroxyacetophenone (liquid film, NaCl plates)

Worked Problem 3.1

Q   The C–H stretching vibrations in dimethyl sulfoxide (Figure 3.18) occur at 2997 and 2909 cm^{-1}. Using Hooke's law, and assuming that the C–H bond is the same strength as the C–D bond (same force constant, k), calculate the stretching frequencies of the C–D bonds in dimethyl sulfoxide-d_6.

A   Hooke's law: $\bar{v} = \dfrac{1}{2\pi c}\sqrt{k/\mu}$, where $\mu = (m_1 \times m_2)/(m_1 + m_2)$

$$\therefore \bar{v}_{CH} = \frac{1}{2\pi c}\sqrt{k_{CH}/\mu_{CH}} \text{ and } \bar{v}_{CD} = \frac{1}{2\pi c}\sqrt{k_{CD}/\mu_{CD}}$$

Since $k_{CH} = k_{CD}$, then:

$$\bar{v}_{CD}/\bar{v}_{CH} = \sqrt{\mu_{CH}/\mu_{CD}}$$

$$\therefore \bar{v}_{CD} = \bar{v}_{CH}\sqrt{\mu_{CH}/\mu_{CD}}$$

Thus $\bar{v}_{CD} = 2997\sqrt{0.92/1.71} = 2198 \text{ cm}^{-1}$ (actual 2250 cm^{-1}) and $\bar{v}_{CD} = 2909\sqrt{0.92/1.71} = 2134 \text{ cm}^{-1}$ (actual 2124 cm^{-1})

You will find the actual values in Figure 3.21, and will also see that the replacement of H by D results in a slight decrease in the stretching frequency of the S=O bond. Remember that Hooke's law is really only applicable to a diatomic molecule, but we are using it to try to predict the stretching frequencies of something much more complex. It is little wonder that the calculated values do not quite match the actual values (although they are not bad!) and that the value for the S=O stretch is also affected.

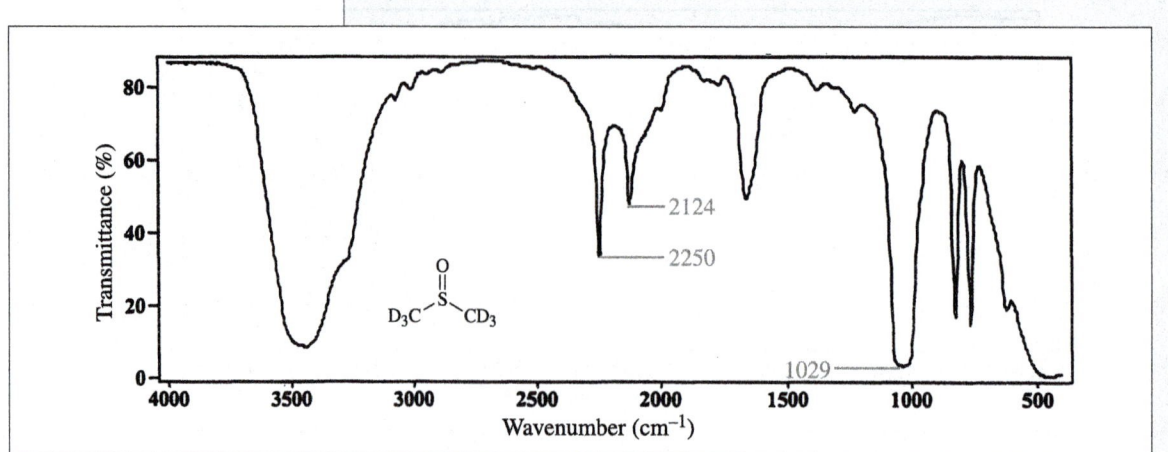

Figure 3.21   IR spectrum of dimethyl sulfoxide-d_6 (liquid film, NaCl plates)

Summary of Key Points

1.   IR spectroscopy involves the study of transitions between the vibrational energy levels of a molecule and the interaction of the oscillating electric vector of the IR light with the oscillating dipole moment of the molecule.

2.   The frequency of the vibration can be described by Hooke's law, which describes the relationship between the wavenumber ($\bar{v}$), the strength of the bond (force constant, k) and the reduced mass μ:

$$v = \frac{1}{2\pi}\sqrt{\frac{k}{\mu}} \quad \text{or} \quad \bar{v} = \frac{1}{2\pi c}\sqrt{\frac{k}{\mu}}$$

3.   Some IR absorptions can be attributed to the vibration of individual bonds, and these can be used to identify functional groups present in a molecule. These characteristic group vibrations can be divided into five regions within the IR spectrum:

4000–2300 cm^{-1}	Vibrations of single bonds to H (C–H, O–H, N–H and S–H)
2300–1850 cm^{-1}	Vibrations of triple and cumulative bonds
1850–1500 cm^{-1}	Vibrations of C=X
1500–1000 cm^{-1}	Fingerprint region
1000–666 cm^{-1}	Unsaturated C–H bending

4. The position of an IR band is affected by many factors, including hydrogen bonding and conjugation, which both result in a lowering of the stretching frequency (wavenumber) for a vibration, and ring strain, which results in an increase in the stretching frequency (wavenumber). Intramolecular hydrogen bonding is independent of concentration, while intermolecular hydrogen bonding is dependent upon concentration.

5. The $1600–1000$ cm^{-1} region is known as the fingerprint region, and the vibrations which occur in this region usually involve the whole molecular skeleton. The fingerprint region is virtually unique to a given molecule, so two unknown samples which are believed to be the same should have identical absorptions in this region.

6. The $1600–1000$ cm^{-1} region corresponds to the bending vibrations of C–H bonds in unsaturated systems, and can often be used to determine the substitution pattern of aromatic rings.

Problems

3.1. Unknown **A** has a molecular formula of $C_8H_9NO_3$. Calculate the number of double bond equivalents and, using the IR spectrum in Figure 3.22, suggest functional groups that might be present.

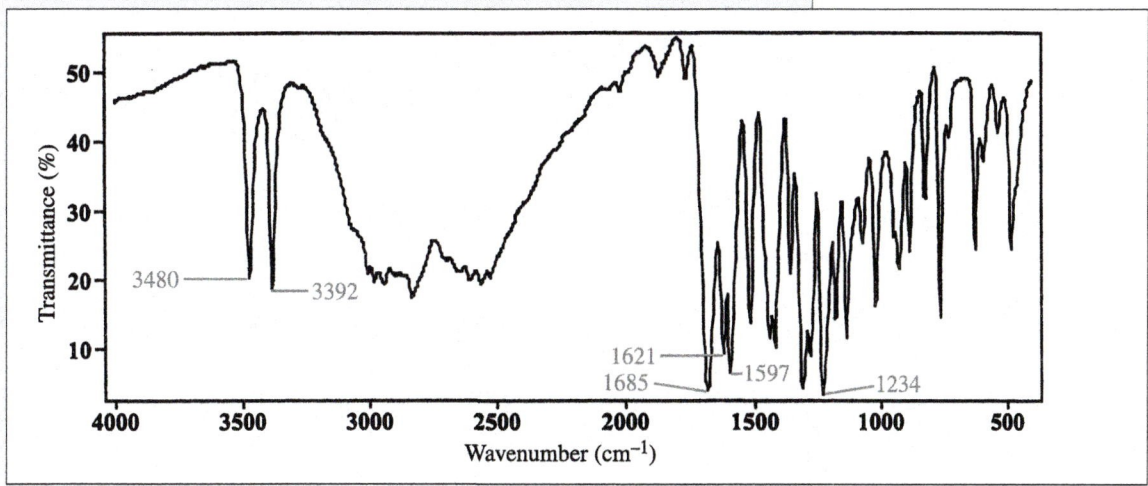

Figure 3.22 IR spectrum of unknown **A** (KBr disc)

3.2. Unknown **B** has a molecular formula of $C_{11}H_{13}NO_3$. Calculate the number of double bond equivalents and, using the IR spectrum in Figure 3.23, suggest functional groups that might be present.

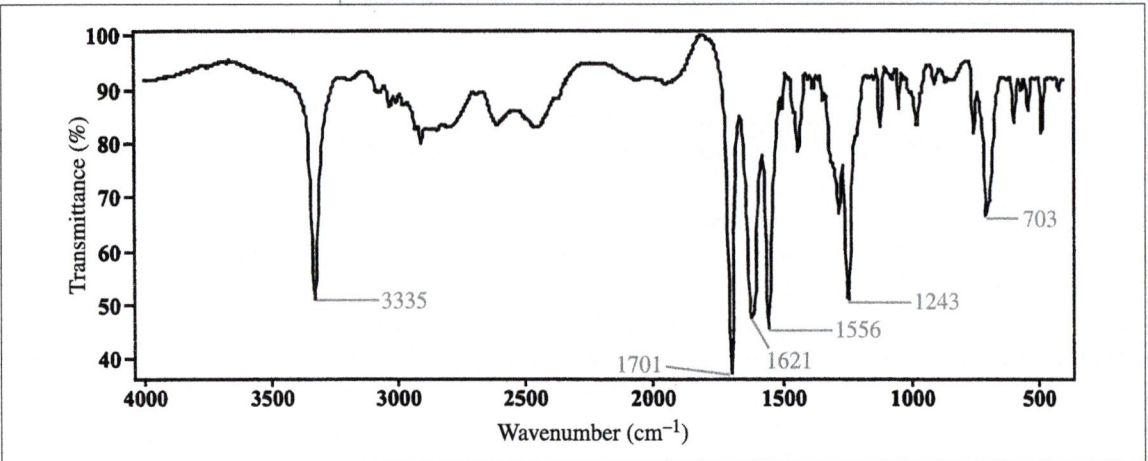

Figure 3.23 IR spectrum of unknown **B** (KBr disc)

3.3. From an analysis of the IR spectrum of unknown **C** (Figure 3.24), suggest functional groups that may be present.

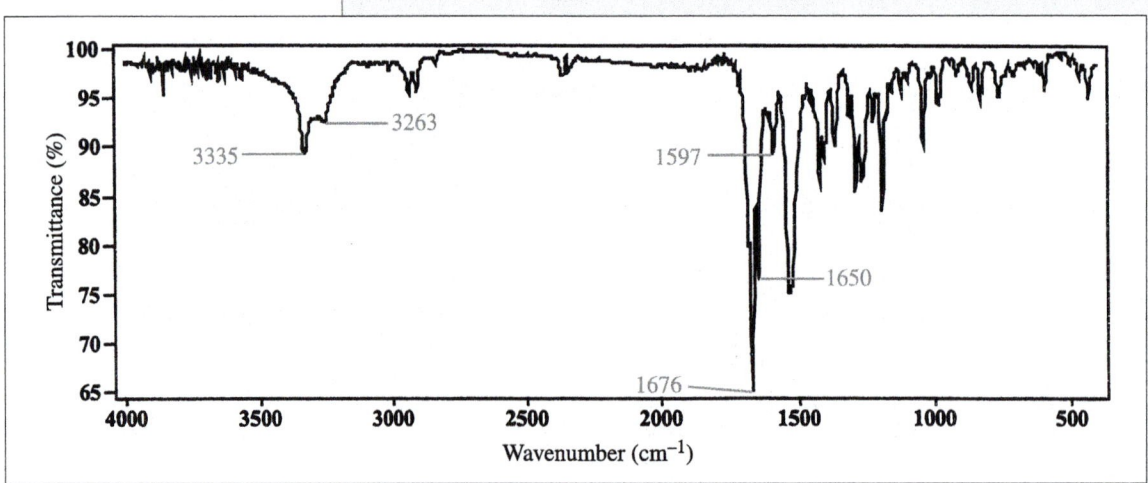

Figure 3.24 IR spectrum of unknown **C** (KBr disc)

3.4. The IR spectra of a pair of isomers, **D** and **E**, of molecular formula $C_{10}H_{16}O$ are given in Figures 3.25 and 3.26. Assign these spectra to the structures shown below.

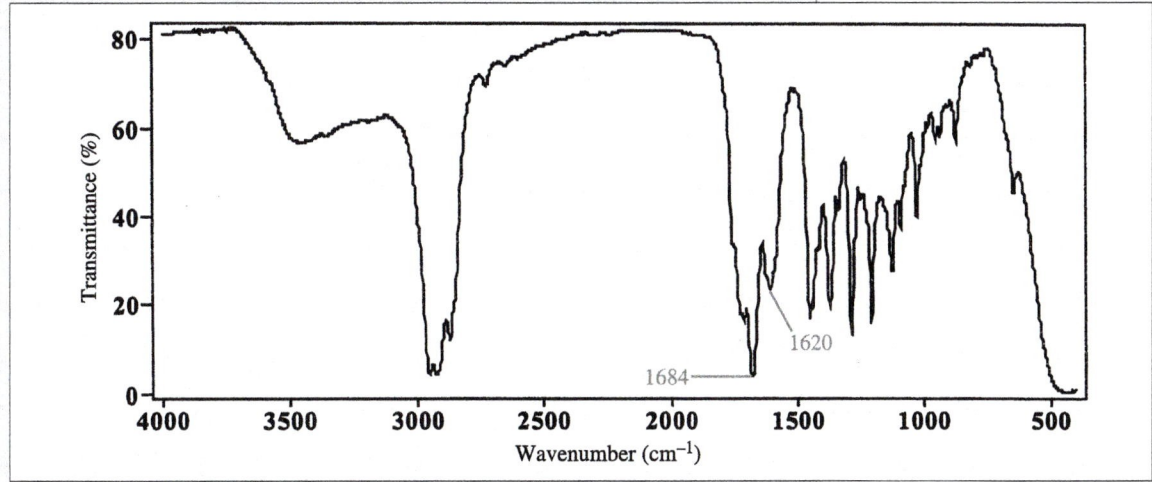

Figure 3.25 IR spectrum of unknown **D** or **E**

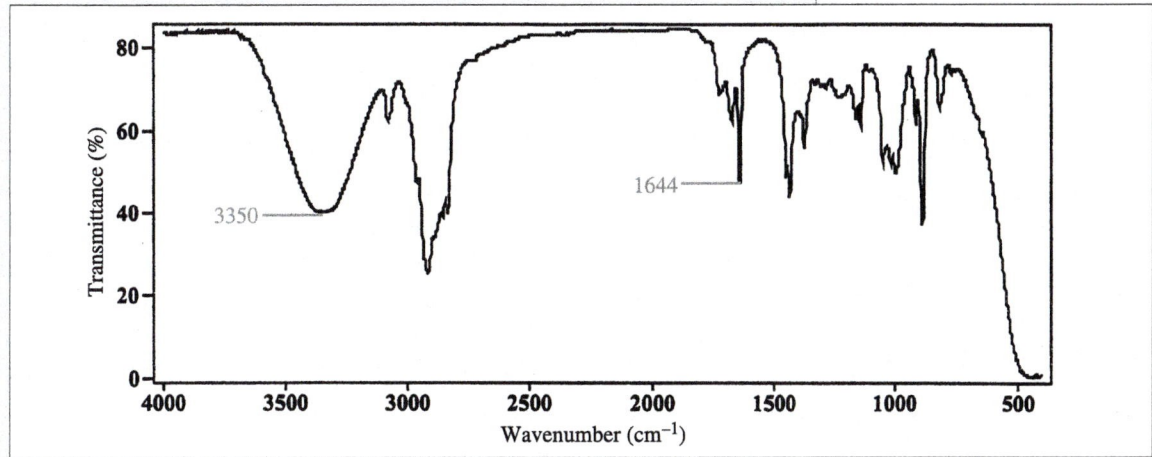

Figure 3.26 IR spectrum of unknown **D** or **E**

3.5. Predict the main absorption bands in the IR spectra of isomers **F**, **G** and **H**.

F   G   H

4

Nuclear Magnetic Resonance Spectroscopy

Aims

This chapter describes the uses of nuclear magnetic resonance (NMR) spectroscopy in organic chemistry. By the end of the chapter you should:

- Understand the origin of the NMR effect, chemical shifts and coupling
- Be able to identify simple organic molecules from their ^{1}H NMR spectra through the interpretation of their integrals (the number of hydrogens in each environment), chemical shifts (the effect of functional groups on the electron density of the molecule) and coupling patterns (the number of spin active nuclei on adjacent atoms)
- Be able to predict peak splitting patterns and estimate coupling constants for protons in a range of molecular environments
- Be able to predict the chemical shift of carbon atoms in the ^{13}C NMR spectra of simple molecules
- Be familiar with the use of various ^{13}C assignment techniques (DEPT, off resonance, HMBC and HMQC)

4.1 Instrumentation

Although the main aim of this book is to provide an aid to the interpretation of organic spectroscopic data, it is difficult to do this without some knowledge of the instrumentation used to carry out the data acquisition. In order to obtain an NMR spectrum, the sample (typically 5–20 mg, depending upon the experiment required) is dissolved in about 0.75 cm^3 of a solvent (Box 4.1), placed in an NMR tube and lowered into the magnetic field of the NMR spectrometer (Figure 4.1).

Box 4.1   Chemical Shifts for Common NMR Solvents

Note that δ_H is the chemical shift for protons in the residual protonated solvent (δ_C is that for the deuterated solvent). The intensities of the peaks in the multiplets will not be the same as for the same multiplets in ^1H spectra as the coupling is to D ($\equiv {}^2H$), a spin $I = 1$ nucleus

Solvent	δ_H	δ_C
$CDCl_3$	7.25 (singlet)	77.0 (triplet)
DMSO-d_6 (CD_3SOCD_3)	2.50 (quintet)	40.4 (septet)
Acetone-d_6 (CD_3COCD_3)	2.05 (quintet)	30.5 (septet), 205.4 (singlet)
Acetonitrile-d_3 (CD_3CN)	1.95 (quintet)	1.2 (septet), 118.0 (singlet)
Methanol-d_4 (CD_3OD)	3.35 (quintet), 4.80 (singlet)	49.0 (septet)
D_2O	4.70 (singlet)	–

4.1

The NMR solvent will obviously be in vast excess over the sample, so ideally should not contain any of the nuclei being investigated in the experiment since these will "swamp" the spectrum. For this reason, fully deuterated solvents, such as deuterochloroform (deuterotrichloromethane, $CDCl_3$) or dimethyl sulfoxide-d_6 (CD_3SOCD_3, **4.1**), are used in proton (^1H) NMR spectroscopy, and are also commonly employed in carbon (^{13}C), fluorine (^{19}F) and phosphorus (^{31}P) NMR spectroscopy. These, or other, fully deuterated solvents often contain trace amounts of the protiated solvent, *e.g.* $CHCl_3$ in $CDCl_3$, owing to incomplete isotopic labelling in their preparation; this is actually quite useful since the peaks for these solvents have well defined chemical shifts (see Section 4.3) and can be used to reference the spectrum. In ^{13}C NMR spectroscopy we have to accept that the best solvent for an organic molecule will normally itself be organic, and so will contain carbon atoms, and the largest peak in the ^{13}C spectrum will therefore be that for the solvent.

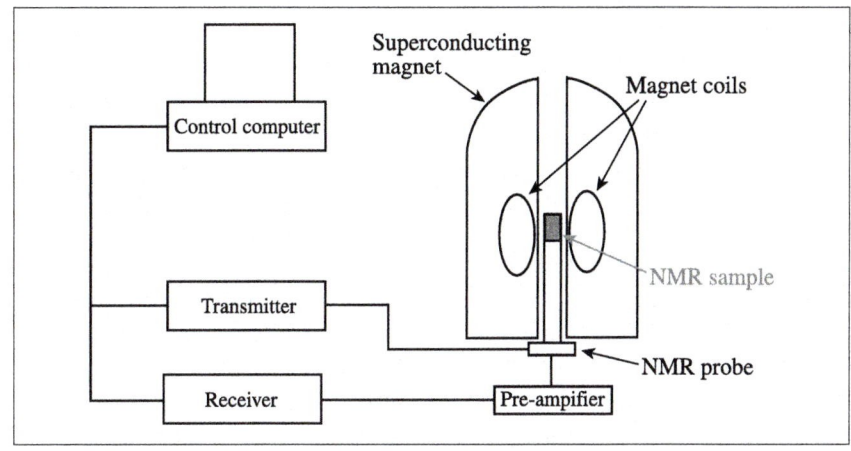

Figure 4.1 Diagram of an NMR spectrometer

Although modern NMR spectrometers appear to be very complex, their operation is relatively straightforward. The various components of an NMR spectrometer are shown in Figure 4.1, and dealing with each of these in turn we can build up a fairly simple picture of how an NMR spectrometer actually operates. The most important (and expensive) part of any modern NMR spectrometer is the super-conducting magnet (SCM), in which an electric current in a coil, which is made from a material which has no electrical resistance at very low temperature ($\sim$4 K or -269 °C), generates the magnetic field. As we will discover later, the strength of the magnetic field used to carry out an NMR experiment determines both the signal-to-noise ratio and the resolution of the data acquired. Indeed, the magnet is such an integral part of any NMR instrument that we usually describe a spectrometer by the frequency at which protons in the magnetic field absorb energy; a 300 MHz instrument is therefore one in which protons undergo their NMR transition at 300 MHz. A key practical consideration, which leads to the major running cost of any NMR instrument, is the need to hold the coil of the magnet at 4 K, since this is currently the highest temperature at which NMR magnets can be made to superconduct.

The only coolant that can be used in such systems is liquid helium, which is expensive to produce and has a low heat capacity.

As we have mentioned above, a critical parameter in determining the signal-to-noise and resolution of our NMR data is the strength of the magnetic field we are using. Another important consideration is the magnetic field homogeneity; the better the field homogeneity of our NMR magnet, the longer the NMR signal will persist and the better the quality of the acquired NMR data. The process of adjusting the homogeneity of the magnetic field is called "shimming" and, together with spinning the NMR tube (at 15–20 Hz), usually produces a suitably homogeneous magnetic field.

The homogeneity of the magnetic field is a measure of how parallel the lines of force of the magnet are: the more parallel, the better the field homogeneity.

When the sample is loaded into the magnet, it ends up inside the sample coil within the NMR probe, the function of which is to transmit radiation (transmitter) to the sample and also to receive the NMR signal (receiver). A great deal of effort has gone into the design of modern NMR probes to make them as efficient as possible in terms of the signal-to-noise ratio, through having the shortest distance possible between the sample and the receiver coil. In addition to being able to spin the sample within the coil, there are also variable-temperature units built into most NMR probes so that, if NMR data are required at different temperatures, a heated or cooled stream of gas can easily be passed over the NMR sample.

Nowadays, virtually all NMR spectra are obtained on Fourier transform (FT) spectrometers. An FT experiment is carried out by placing the solution of the sample in a magnetic field and by irradiating it with a short pulse (broad spectrum) of radiofrequency radiation. The effect of this irradiation over such a broad range is that all the nuclei in the sample are excited at the same time. The NMR spectrometer then detects the radiation given off by the sample as the nuclei "relax" back to their lower energy state, and one way in which they can do this is by transferring energy to other, nearby, nuclei with spin (we will learn a little more about this later).

The nuclei in the sample that are present in differing chemical environments (see Section 4.3 on chemical shift) give off radiation with characteristic frequencies (sine waves with different frequencies) and the net result is interfering wave patterns of differing frequency, forming an interferogram or free induction decay (FID) – a plot of signal intensity against time. In order to obtain the spectrum in the form we are familiar with, this FID is Fourier transformed – a mathematical process which, thankfully, we do not need to understand – to give a plot in which the x-axis represents the particular frequency at which a transition occurs and the y-axis represents the intensity associated with this transition.

The improvement in the signal-to-noise ratio (S/N) is proportional to the square root of the ratio of the number of scans. For example, a spectrum acquired with 64 scans will have an improved S/N of $\sqrt{(64/16)} = \sqrt{4} = 2$ over one acquired with 16 scans.

One of the main advantages of FT spectrometers is that, since the FID is in digital form, we can repeat the excitation/detection process a number of times and all the resulting "scans" can be added and the FT performed on the resultant FID. In this way, we can improve the signal-to-noise ratio and can detect nuclei which are not very abundant (*e.g.* ^{13}C) or have low sensitivity to NMR (see Section 4.2). These nuclei could not have been detected on the older continuous wave instruments, as the spectrum was the result of a single scan, obtained as one of the frequency or magnetic field were varied while keeping the other constant.

4.2   Origin of the NMR Effect

To explain the origin of the NMR effect, we have first to think about why something has a magnetic field associated with it. Consider a simple compass, the needle of which is a small magnet and is therefore affected by the presence of other magnetic fields (*i.e.* it moves in the presence of, or interacts with, other magnetic fields). The Earth has associated with it its own small but nonetheless significant magnetic field and a compass will interact with this and will always point North. We now have a means of detecting magnetic fields: take a compass near the suspected source and watch for an interaction, *i.e.* some movement of the needle. If we attempt this simple experiment near a wire carrying an electric current, we find that we can detect the presence of a magnetic field and this gives us a clue to one particular source of a magnetic field: all moving charges have associated with them not only an electric field but also a magnetic field.

What has this got to do with NMR? An atomic nucleus has a positive charge associated with it (which is why the negatively charged electrons are maintained in their orbit) so we have one component of what is required for a magnetic field – a charge. In addition, certain nuclei, with odd mass or atomic numbers, spin because their spin quantum number I is not zero ($I \neq 0$) (Box 4.2), and so we would expect to find a magnetic field associated with these "nuclear magnets". If we could find a compass with a small enough needle (*i.e.* one that was of the order of the size of an atomic nucleus, $\sim 10^{-10}$ m), we would find that it would interact with nuclei which had $I \neq 0$. Of course, this is not just a one-way effect, as the atomic nucleus not only affects magnetic fields brought near it but is itself affected by such fields. Rather annoyingly, some of the most common nuclei found in organic molecules (^{12}C, ^{16}O and ^{32}S) have both even mass and atomic numbers (Box 4.3) and so have a spin quantum number of 0, and can never exhibit an NMR effect.

Box 4.2   Atomic Nuclei Orientation

Atomic nuclei with spin $I \neq 0$, when placed into a magnetic field, cannot adopt any alignment but exhibit quantization, with their orientation depending upon their spin quantum number, I. The nuclei can adopt $(2I+1)$ orientations in the field, with m_I (magnetic quantum number) values of $-I, -I+1, \ldots, I$. Transitions with $\Delta m_I = \pm 1$ are allowed.

Box 4.3   Atomic Nuclei and Spin

Atomic number	Mass number	Spin, I	Examples (natural abundance/%)
Even	Even	0	^{12}C (98.89), ^{16}O (99.76), ^{32}S (95.04)
Odd/even	Odd	½, 3/2, 5/2, etc.	^{1}H (99.99), ^{13}C (1.11), ^{19}F (100), ^{31}P (100) all $I = ½$
Odd	Even	1, 2, 3, etc.	^{14}N (99.64), ^{2}H (0.015) both $I = 1$

When a magnetic field (the symbol for which is B) is applied to nuclei with spin quantum number $I \neq 0$, the nuclei can adopt $(2I + 1)$ orientations in the field (Box 4.2), and the NMR selection rule states that transitions with $\Delta m_I = \pm 1$ are allowed. For a nucleus with spin $I = ½$ (such as ^{1}H, ^{13}C, ^{19}F and ^{31}P), there are therefore two possible arrangements: either aligned "with" (lowest energy state, α, $m_I = +½$) or aligned "against" (highest energy state, β, $m_I = -½$) the applied field, B (Figure 4.2).

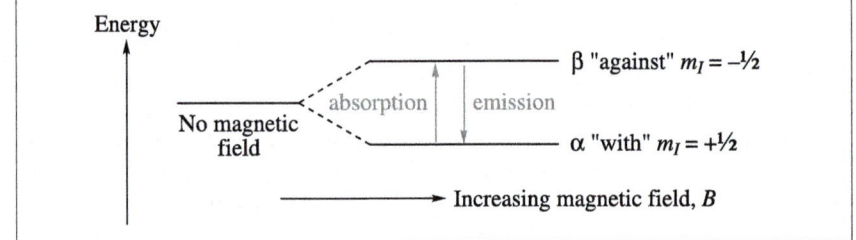

Figure 4.2   Allowed orientations for a nucleus with spin $I = ½$ in a magnetic field

From basic spectroscopic theory, $E = h\nu$ (see Chapter 1), and we know that when there are two differing energy levels we can observe allowed transitions between them by either monitoring the absorption or emission of electromagnetic radiation – hence the origin of the NMR effect. The NMR experiment involves moving atomic nuclei from the lower energy, α, state ("with" or parallel alignment) to the higher energy, β, state ("against" or anti-parallel alignment) and then monitoring the effect this has on the sample of nuclei being observed.

As you can imagine, the size of the magnetic field generated by a single atomic nucleus is incredibly small and the size of the interaction of such a small nuclear magnet with even the strongest magnetic fields (*i.e.* those generated by modern superconducting magnets) is, therefore, also incredibly small. We would expect therefore to find that NMR transitions are induced by very low frequency radiation and this indeed turns out to be the case. NMR transitions are of such a low energy that they occur in the radiofrequency (low energy) end of the electromagnetic spectrum, and the relationship between the magnetic field and the frequency of the transition is given in equation (4.1):

$$v = \gamma B_{\text{eff}}/2\pi \qquad (4.1)$$

where v is the frequency at which the NMR transition occurs, γ is the magnetogyric ratio and B_{eff} is the field experienced by the nucleus.

The energy difference between the possible alignments of the nuclear magnetic dipole is thus dependent on the strength of B: the stronger the applied field, the greater the energy gap and the greater the frequency (Figure 4.2). When there is no applied field, there is no energy difference between the possible alignments of the nuclei (the energy levels are said to be degenerate) and the NMR effect is not observed. Even in the magnetic fields produced by the strongest superconducting magnets the difference in energy between the possible alignments of the nuclear magnet is very small. This small energy difference gives rise to another feature of NMR spectroscopy: the poor signal-to-noise ratio. The population difference between quantized energy levels is proportional to the energy difference between them and, because of the small energy gap between the energy levels of the nuclear dipole in a magnetic field, the Boltzmann distribution for a nucleus with spin $I = \frac{1}{2}$ at room temperature is very nearly 50:50, *i.e.* approximately 50% of the nuclei will be aligned "with" and 50% of the nuclei will be aligned "against" the magnetic field. Therefore we always attempt to conduct NMR experiments at the highest frequency possible; since the value for γ is constant for a given nucleus, we do this by making the strength of the main magnetic field (B) as large as we can. This is one of the main reasons why chemists pay very large sums of money for superconducting NMR magnets with the strongest field possible (of course, there is also an element of ego here!).

4.3 Chemical Shift

The first thing we need to consider when discussing the NMR effect is why equation (4.1) contains the term B_{eff} rather than B. This is called the chemical shift effect, and to understand this key aspect of NMR spectra we need to consider what happens when we place our NMR sample in a magnetic field. The atomic nuclei in our sample adopt one of a number of possible alignments, depending upon their spin. We also, however, have to

v is the frequency of the radiation required to induce a particular nucleus to undergo its NMR transition (*i.e.* in changing its alignment from being "with" to "against" the main magnetic field) and has the units Hz (or s^{-1}); B_{eff} is the symbol used for the strength of the magnetic field experienced by the nucleus and has the units T (Tesla) (the applied magnetic field, B, is actually modified by the molecular environment of the nucleus being observed to give B_{eff}: see Section 4.3 on chemical shift; γ is the magnetogyric ratio (units of s^{-1} T^{-1}), which is a measure of the size of the magnetic field, or magnetic moment, generated by a particular nuclear species. All ^{1}H nuclei (the nucleus most commonly studied using NMR spectroscopy) have one value for the magnetogyric ratio and all other NMR-active nuclei (*e.g.* ^{13}C, ^{19}F and ^{31}P) also have their own respective values for this constant, which in all cases is smaller than that for ^{1}H nuclei.

In actual fact, there is an excess of only about 1 in 10^5 nuclei in the lower energy level so that, when conducting an NMR experiment, we are working with just 1 in 10^5 of the NMR-active nuclei!

take into account the fact that the atoms in our NMR sample do not just contain one charged species but also contain a second (electrons). What effect does the main field have on the electrons within the atoms in our NMR sample? Because of their charge, electrons also interact with the magnetic field and begin to circulate around the nucleus. As we have already discovered, moving charges generate magnetic fields and, as happens for the nucleus itself, the atomic electrons produce their own magnetic field when the NMR sample is placed into a magnet. In the case of the atomic electrons, the field produced is such that it opposes the main field (Figure 4.3) and therefore has the effect of decreasing B (the applied field) at the nucleus. This decrease in the strength of the main field is tiny (and can be measured in parts per million when compared to the main field), but is nonetheless detectable when the NMR experiment is performed. The size of this chemical shift effect is obviously dependent upon the electron density surrounding the nucleus: we would expect a nucleus surrounded by high electron density to be most affected by the opposing magnetic field generated by the electrons, and conversely one with a low electron density to show a lesser chemical shift effect. NMR is therefore uniquely sensitive to the molecular environment of the nuclei being observed.

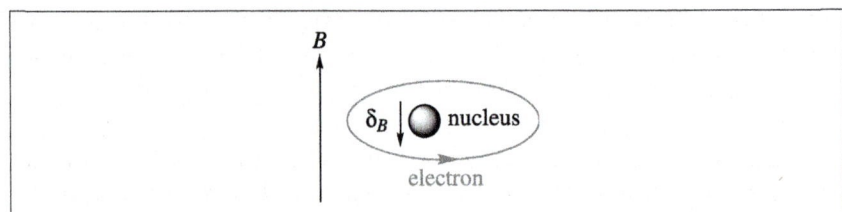

Figure 4.3 Interaction of the electron magnetic field with the main magnetic field

This is, in fact, what we find when we do an NMR experiment: if an atom is attached to an electron-donating substituent, the result is an increase in the electron density surrounding the nucleus, and a *decrease* in B_{eff}. This modification of the main field (B) by the surrounding electrons is often called shielding (as the nucleus is effectively being "shielded" from the main field by the magnetic field generated by the electrons); the greater the electron density, the higher the shielding (and lower the frequency for the transition). The term used to describe the situation where an electron-withdrawing group causes a reduction in the electron density and, hence, an *increase* in B_{eff}, is deshielding. A deshielded nucleus thus resonates (another commonly used term to describe the frequency at which an NMR transition occurs) at a *higher frequency* than would have been expected. If B_{eff} is greater for a more deshielded nucleus, then, from equation (4.1), the frequency at which the NMR transition occurs is clearly higher. The end of an NMR spectrum where deshielded nuclei appear (high parts per million, or ppm, values) is termed the high-frequency (or, more commonly, the low-field) end of the spectrum, while shielded nuclei appear at the low-frequency (high-field) end.

The really important aspect to all of this is that nuclei in similar chemical environments exhibit similar chemical shifts. Thus protons (^{1}H nuclei) attached to a carbon atom bonded to oxygen, H–C–O, show a characteristic chemical shift (3.5–4.5 ppm), while protons attached to a carbon atom bonded to nitrogen, H–C–N, have a different chemical shift range (2.5–3.5 ppm) and, since the carbon is attached to the less electronegative N atom, resonate at lower frequency. We can therefore use chemical shifts to our great advantage when interpreting NMR spectra.

The usual units used to measure the chemical shift effect are parts per million (ppm). As was stated above, this is due to the fact the chemical shift effect is very small when compared to the main field. NMR can be carried out at many different magnetic field strengths, depending on the strength of the magnets used for the experiment. We know from equation (4.1) that the frequency of a transition is directly proportional to the strength of the magnetic field, B. If we were to use frequency as our units of measurement of chemical shift, we would find ourselves having to issue tables of chemical shift values for each possible value of B. We remove this unwanted magnetic field (B) dependence by defining chemical shift (δ) to be the ratio of the frequency at which an NMR transition occurs to the strength of the main field as measured by the frequency at which protons resonate (equation 4.2):

$$\delta = \frac{\nu_{sample} - \nu_{reference}}{\nu_{spectrometer}} \times 10^6 \text{ (ppm)} \qquad (4.2)$$

In $CDCl_3$ or DMSO-d_6 we normally choose tetramethylsilane (TMS), $(CH_3)_4Si$, as our reference, and this is given a reference value of 0 Hz so that, in the above example, the signals which occur at 100 Hz and 500 Hz, at magnetic fields of 100 MHz and 500 MHz respectively, have a chemical shift of δ 1.0 in both cases.

> A transition that occurs at 100 Hz on a 100 MHz NMR spectrometer will occur at 500 Hz on a 500 MHz instrument, although the ^{1}H giving rise to the signal is exactly the same.

4.4 ^{1}H NMR Spectroscopy

4.4.1 ^{1}H Chemical Shifts

A ^{1}H NMR spectrum consists of a plot of chemical shift against intensity and, since *the area under each peak in a ^{1}H NMR spectrum is proportional to the number of ^{1}H nuclei giving rise to the peak*, we integrate the spectrum to give the number of protons in each molecular environment (Figure 4.4, red trace). As mentioned previously, chemical shift values are distinctive for the chemical environment of a particular nucleus. Thus, similar chemical environments give rise to signals with similar chemical shifts, and these are often tabulated to allow us to use chemical shift as a starting point for our interpretation of NMR spectra. The ^{1}H chemical shifts for some of the most common functional groups are given in Box 4.4, but much more extensive lists are available elsewhere. The first thing you will notice about the chemical shifts in Box 4.4 is that the range is rather small (typically δ 0–15), and this is due to the fact that the chemical shift of protons is influenced only by the 1s electron. For all other nuclei (which have p-electrons), much greater effects on the chemical shift are seen and, for example, the usual chemical shift range for ^{13}C is δ 0–220. You will also notice from Box 4.4 that the more electronegative the atom which the carbon bearing the proton is attached to, the more deshielded the proton (the higher the frequency). Thus, H–C–O have higher chemical shifts than H–C–N, which have higher chemical shifts than H–C–C. ^{1}H Chemical shifts are thus a measure of electron density and the more electronegative the group attached to the carbon bearing the proton, the higher the chemical shift (δ) of that proton.

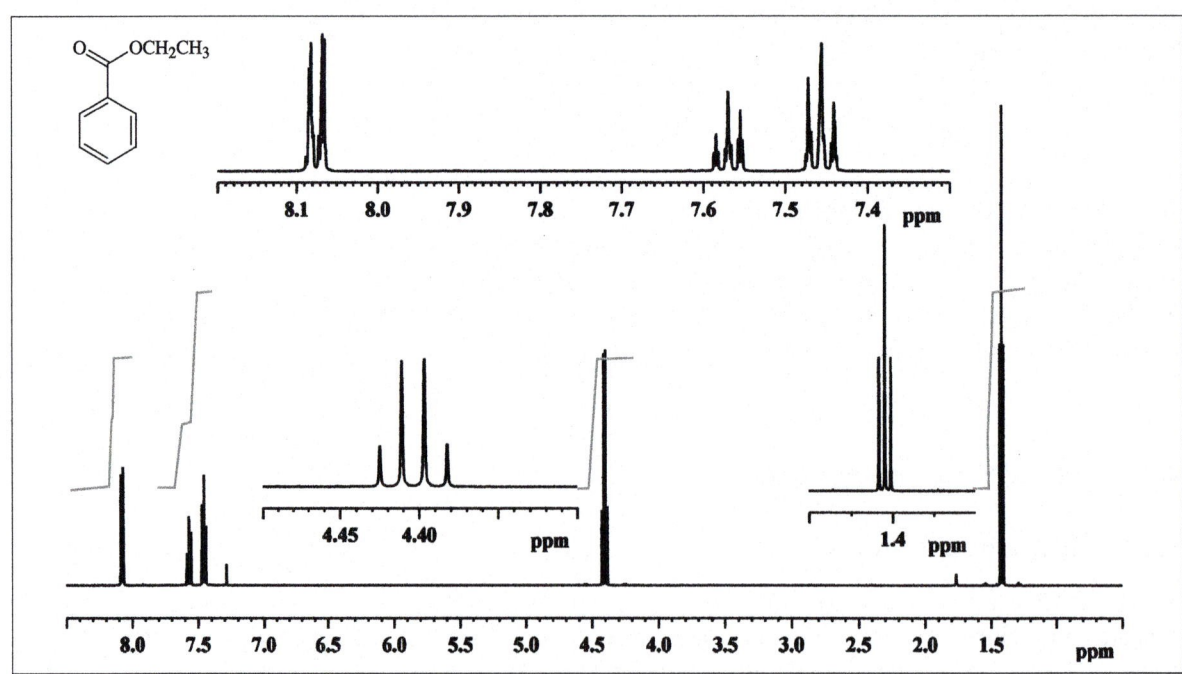

Figure 4.4   500 MHz ^{1}H NMR spectrum of ethyl benzoate in CDCl$_3$ (δ_H 7.25)

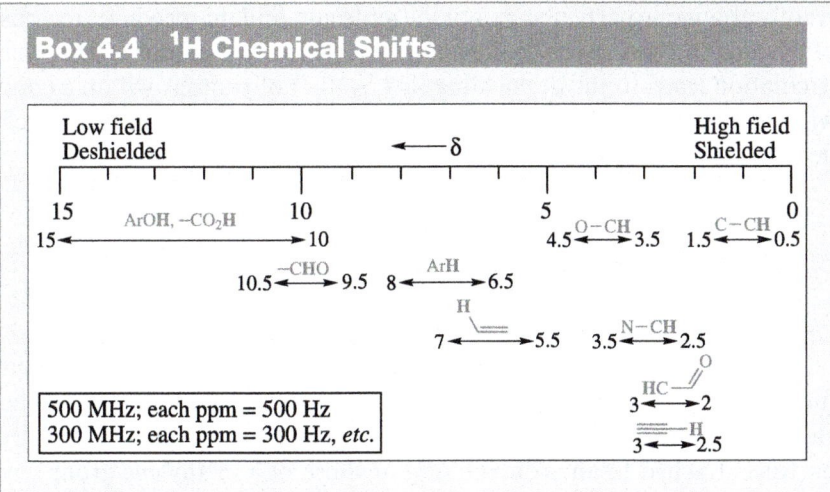

One particularly noticeable feature of Box 4.4 is the effect of aromaticity on the chemical shift: you can see that protons attached to an aromatic ring have chemical shifts in the region δ 6.5–8. If the case of a benzene ring is considered, then this effect is very clearly illustrated (Figure 4.5): because of the presence of the delocalized electrons that occur above and below the aromatic ring, when the ring is placed in a magnetic field a ring current is set up in which the aromatic electrons of the ring circulate in the planes above and below the ring. As we saw earlier, moving charges generate magnetic fields and, in this case, there is a reinforcement of the main field to produce a higher than expected B_{eff} for an aromatic system. Thus, the protons of a benzene ring occur at a lower field (higher ppm value) than would have been expected in a molecule such as this – that is, they are deshielded.

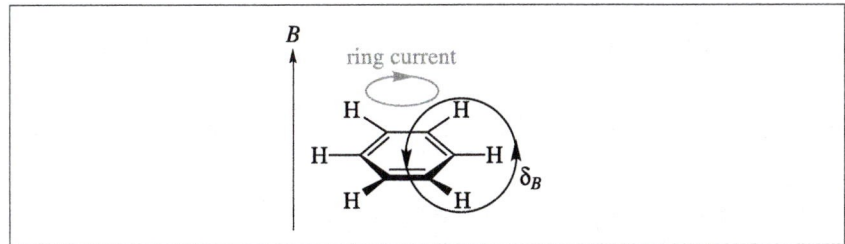

Figure 4.5 Aromaticity and chemical shift

Similar effects are seen in functional groups which contain multiple bonds and so have electrons which can circulate in an applied magnetic field. In some cases, this circulation of the π-electrons happens more

Anisotropy: where properties differ according to the direction of measurement.

readily about one particular axis of the molecule, and this produces regions of "magnetic anisotropy". For example, in a carbonyl group, –CHO, this circulation leads to the deshielding (δ 9.5–10.5) of protons within a cone whose axis is along the carbonyl bond, while protons in the molecule which lie close to this group, but outside this cone, are shielded (Figure 4.6).

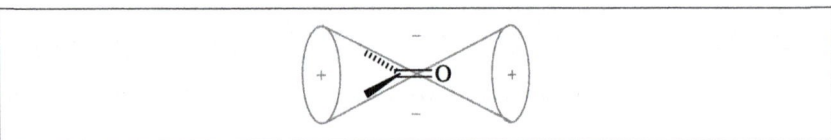

Figure 4.6 Shielding and deshielding for a carbonyl group (+ represents deshielding, – shielding)

The values of chemical shifts for protons attached to different functional groups are reasonably constant, so that we can predict the chemical shift of protons in various molecular environments (especially protons attached to an aromatic ring or those of a methylene group) by using some empirical rules (Boxes 4.5 and 4.6).

Box 4.5   Shoolery's Rules for X^1–CH_2–X^2 and X^1–CHX^2–X^3

$$\delta_H = 0.23 + \sum \Delta_X$$

X	Δ_X
H	0.18
Alkyl (Me, Et, *etc.*)	0.50
C=C	1.25
C≡C	1.38
CO_2R (acids and esters)	1.56
NR_2 (primary, secondary, or tertiary amines)	1.58
COR (aldehydes, R = H, and ketones)	1.69
C≡N	1.69
I	1.82
Ph	1.85
Br	2.36
Cl	2.53
OH	2.55
F	3.25

This table was compiled using chemical shifts taken from spectra in the SDBS (http://www.aist.go.jp/RIODB/SDBS/menu-e.html) database. The data are most accurate for methylene (CH_2) groups and can also be used to estimate the chemical shifts of methyl (CH_3) groups from X^1–CH_2–X^2, with X^1 = H.

Box 4.6   Substituent Effects on the Chemical Shifts of Aromatic Protons

$$\delta_H = 7.34 + \sum \Delta_X$$

Substituent, X	$\Delta_{X\,ortho}$	$\Delta_{X\,meta}$	$\Delta_{X\,para}$
Me	−0.27	−0.20	−0.29
Ph	0.25	0.10	0.01
CH_2OH	−0.14	−0.14	−0.14
COR	0.49	0.15	0.25
CO_2R	0.75	0.09	0.23
$C=CH_2$	−0.02	−0.10	−0.15
$C\equiv CH$	0.08	−0.10	−0.08
$C\equiv N$	0.27	0.10	0.21
OR	−0.48	−0.09	−0.42
SR	−0.11	−0.10	−0.24
NH_2	−0.70	−0.22	−0.61
NO_2	0.85	0.18	0.31
F	−0.31	−0.03	−0.21
Cl	−0.05	−0.10	−0.15
Br	0.13	−0.15	−0.07
I	0.33	−0.29	−0.04

This table was compiled using data taken from spectra in $CDCl_3$ on the SDBS Website (http://www.aist.go.jp/RIODB/SDBS/menu-e.html): δ_H (benzene, in $CDCl_3$) 7.34.

Example: benzocaine (ethyl 4-aminobenzoate)

H2/H6 are *ortho* to the ester (CO_2R) and *meta* to the amino (NH_2) groups:

$$\therefore \delta_H(H2/H6) = 7.34 + \Delta_{ester}(ortho) + \Delta_{amino}(meta)$$
$$= 7.34 + 0.75 + (−0.22) = 7.87$$

H3/H5 are *ortho* to the amino (NH$_2$) and *meta* to the ester (CO$_2$R) groups:

$$\therefore \delta_{\text{H}}(\text{H3/H5}) = 7.34 + \Delta_{\text{amino}}(ortho) + \Delta_{\text{ester}}(meta)$$
$$= 7.34 + (-0.70) + (0.09) = 6.73$$

In the 90 MHz ^{1}H NMR spectrum of benzocaine in CDCl$_3$, the actual values are δ7.85 and 6.63.

These "exchangeable" peaks can be distinguished readily by adding a drop of D$_2$O to the sample and shaking. This converts the NH or OH to ND or OD and their signals in the ^{1}H NMR spectrum are lost (there will be a corresponding peak in the ^{2}H NMR spectrum, however, and a peak appears at about δ 4.5 in the ^{1}H NMR spectrum owing to the formation of HOD).

Among organic molecules, the most acidic are the carboxylic acids (RCO$_2$H), and so it is no surprise that the proton of the carboxyl group, with very low electron density (acidic), is strongly deshielded (δ 10–15). Other noticeable features of acid protons, and of any NH, SH or OH, is that their NMR peaks are generally broad (owing to rapid exchange, especially with the solvent and any water present, these protons are continually being lost to, and regained from, other molecules), do not couple to other protons (again owing to fast exchange) and have variable chemical shifts (which are dependent upon the solvent, the concentration of the sample and the temperature). The question of the rates of exchange processes and their effect on an NMR spectrum is complex and we do not have space to go into it in depth here, but, as a general rule, anything which happens at a rate greater than once per millisecond (1/1000 s) will lead to broad NMR peaks, while anything which happens more slowly will give rise to discrete peaks for the individual protons which are in exchange. The exchange of acidic protons with the solvent usually happens much more often than once per millisecond and these peaks appear broad and have no coupling.

An amide bond has partial double bond character and so restricted rotation (Figure 4.7).

While considering exchange processes, it is often common to see more peaks than would be expected in the ^{1}H NMR spectra of amides; the reason for this is that amides can exist as rotamers about the amide bond. This phenomenon has already been discussed (in Section 3.3.3) as it leads to a decrease in the CO bond order and a resulting decrease in the stretching frequency for an amide bond vibration. These rotamers are in slow exchange on the NMR timescale, so that the NMR spectrum can appear to be a mixture of two discrete compounds, with an example of this process shown in the spectrum of *N,N*-dimethylformamide (DMF, *N,N*-dimethylmethanamide) (Figures 4.7, 4.8), in which the two methyl groups are clearly not equivalent at 353 K, owing to restricted rotation about the C–N bond. If restricted rotation is suspected of complicating a ^{1}H spectrum, we can re-run the spectrum at higher temperatures. This should overcome the energy barrier to the rotation and, for example, the peaks for the methyl groups in DMF would broaden and coalesce into one peak. At the coalescence temperature (413 K for DMF in DMSO-d_6), the rotation is happening at about the NMR timescale, and we see a characteristically broad peak for the N-methyl groups. At temperatures above the coalescence temperature there is a singlet for all six H atoms of the N-methyl groups, since the C–N bond is in free rotation and the two methyl groups are equivalent. By determining the coalescence temperature (T_c) and by knowing the chemical shift difference (∂v in Hz) of the two peaks which coalesce, we can calculate the free energy of activation for the process using equation (4.3):

$$\Delta G^{\ddagger} = RT_c \left[22.96 + \ln \left(\frac{T_c}{\partial v} \right) \right] \tag{4.3}$$

From Figure 4.8, T_c is 413 K and ∂v can be measured as 48 Hz, which is the separation in Hz between the two methyl peaks at 353 K (in this case) so that:

$$\Delta G^{\ddagger} = 8.314 \times 413 \times \left[22.96 + \ln \left(\frac{413}{48} \right) \right] = 3434 \times [22.96 + \ln (8.6)]$$

$$= 3434 \times [22.96 + 2.15] = 3434 \times 25.1 = 86.2 \text{ kJ mol}^{-1}$$

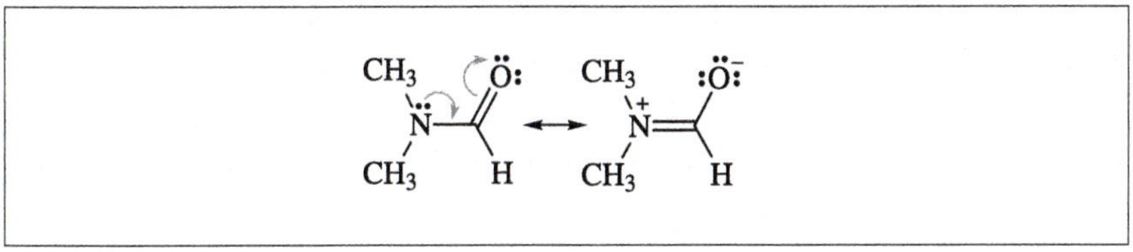

Figure 4.7   Restricted rotation in N,N-dimethylformamide

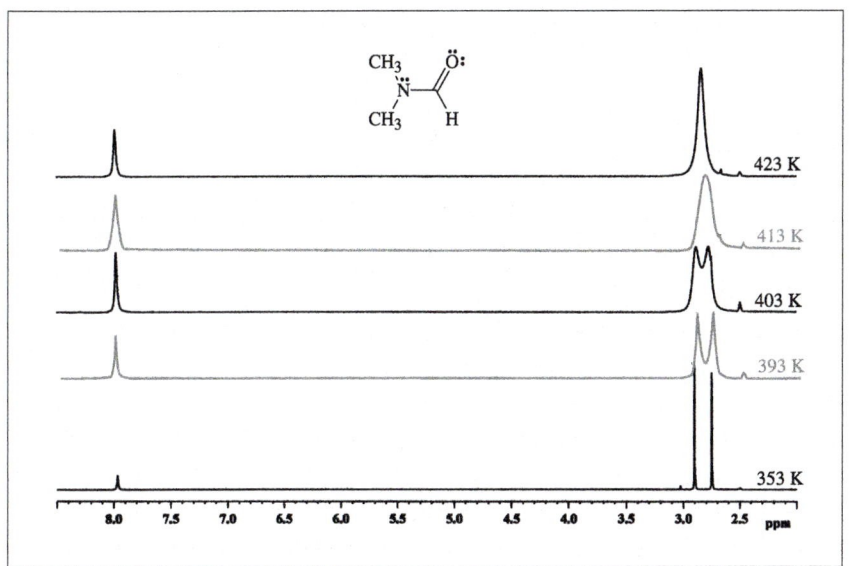

Figure 4.8   Variable-temperature ^{1}H NMR spectra of N,N-dimethylformamide in DMSO-d_6; coalescence temperature, $T_c = 413$ K ($\Delta G^{\ddagger} = 86.2$ kJ mol^{-1})

4.4.2 Spin–Spin Coupling

From what has been said so far, we might expect that single protons, *e.g.* H–CCl$_3$, and groups of chemically equivalent protons, *e.g.* CH$_3$–R, would give rise to single ^{1}H NMR peaks. This does happen, but only if there are no other magnetic nuclei close by in the molecule, *i.e.* for us to obtain a singlet peak for a given proton there must, in general, be no protons on adjacent carbon atoms.

IUPAC is the acronym for the International Union of Pure and Applied Chemistry, the body that oversees and agrees international standards in chemistry, *e.g.* the nomenclature of organic compounds.

At low field (high frequency) in the ^{1}H NMR spectrum of dichloroacetaldehyde diethyl acetal (dichloroethanal diethyl acetal or, to give the IUPAC systematic name, 1,1-dichloro-2,2-diethoxyethane) (Figure 4.9), we can see that there are two doublets (2 sets of 2 lines) at δ 4.63 and δ 5.59 (we will look at the higher field end of this spectrum later). The origin of these lines, known as splitting patterns, is so important that we shall try to explain it in two different ways here, and we will also show you how to predict such splitting patterns in two ways. An understanding of NMR splitting patterns is crucial to the ability to elucidate structures by NMR since the patterns, and the separation between the peaks (the coupling constant, J), give key information about the molecular structure.

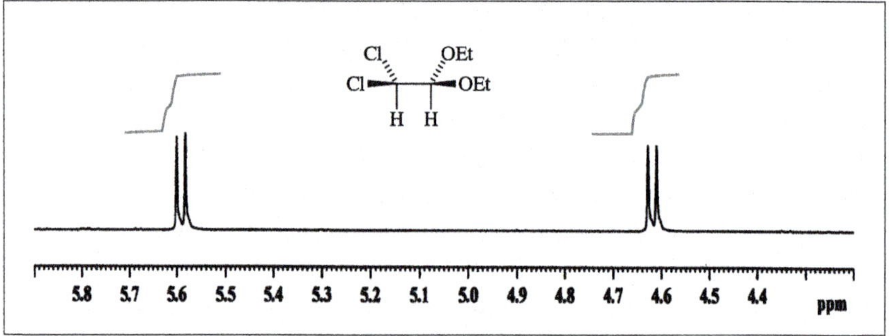

Figure 4.9 Low-field (high-frequency) region of the 300 MHz ^{1}H NMR spectrum of dichloro-acetaldehyde diethyl acetal in CDCl$_3$

Have a look again at the structure of dichloroacetaldehyde diethyl acetal: we can see that three bonds separate the protons marked in bold in this structure. In aliphatic molecules such as this, or in the aliphatic portions of larger molecules, spin–spin coupling is usually restricted to a maximum of three bonds between the atoms which are coupled. Thus these protons couple to one another, and give rise to the doublets at δ 4.63 ($J = 5.6$ Hz) and δ 5.62 ($J = 5.6$ Hz), and there is no interaction between either of these protons with those of the ethyl groups. Our first question should be: how do these doublets arise? There are, of course, many other questions we need to ask, but we will come to them later.

The doublets arise due to spin–spin coupling, which is a magnetic "through-bond" interaction between the two nuclei which are coupled (we will call them H^A and H^X, as this is the convention), brought about by the interactions between local magnetic fields and bonding electrons. The magnitude of the coupling constant, J, depends upon the bonding system between the nuclei, with fewer bonds between the nuclei giving rise to larger values of the coupling constant. A key point we need to know and remember is that *chemically and magnetically equivalent protons do not couple to one another*.

The Origin of Spin–Spin Coupling

Remember that a proton, or any other nucleus with spin $I = \frac{1}{2}$, can have $(2I + 1) = 2$ orientations in an applied magnetic field. The two possible orientations are aligned "with" ($m_I = +\frac{1}{2}$) and "against" ($m_I = -\frac{1}{2}$) the applied field. Remember also that the frequency v of an NMR transition is proportional to the magnetic field experienced by the nucleus, B_{eff}.

Recall equation (4.1):

$$v = \gamma B_{eff}/2\pi$$

The Effect of H^A on H^X (and H^X on H^A)

Consider first H^A, and how the magnetic environment of this proton is influenced by the presence of H^X. If H^X were absent, H^A would experience a magnetic field of B^A. However, in approximately 50% of the molecules in the sample (remember that the population differences between the two energy levels are incredibly small so that the energy levels are almost equally populated), H^X will be aligned "with" the applied magnetic field, B, so that H^A will experience a magnetic field of $B^A + \Delta B^X$ (H^X reinforces the applied field) (Figure 4.10a). Thus, in this 50% of the sample molecules, H^A will absorb at a slightly higher frequency than v_A, and we say that, in this case, H^A is deshielded.

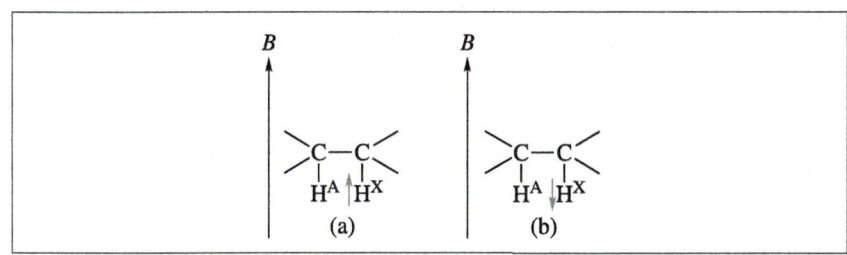

Figure 4.10 In about 50% of the molecules in the sample, H^X will be aligned "with" B (a) and in about 50% H^X will be aligned "against" B (b)

In the other approximately 50% of the molecules in the sample, however, H^X will be aligned "against" B (Figure 4.10b), so that H^A will experience a magnetic field of $B^A - \Delta B^X$ and will thus absorb at a slightly lower frequency than v_A (shielded). Thus, instead of a single line for H^A at v_A, we observe a doublet (two lines of equal intensity), with one line at a slightly higher frequency than v_A and the other at a slightly lower frequency (Figure 4.11). The separation between the lines is known as the coupling constant, J_{AX}, and is always measured in Hz.

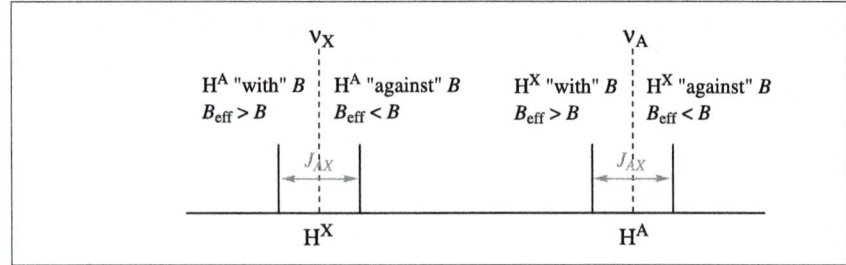

Figure 4.11   Doublets for H^A and H^X

There is an equal and opposite interaction between H^A and H^X which gives a doublet of J_{AX} for H^X. Look back again at Figure 4.9, and you will see that what we predicted is exactly what we observe for the two CH protons of dichloroacetaldehyde diethyl acetal, $Cl_2CHCH(OCH_2CH_3)_2$.

The Energy Level Diagram for Two Interacting Nuclei, H^A and H^X

For a more detailed analysis we need to look at the energy levels for the particular spin system. For a two-spin system ($H^A H^X$) there are four energy levels, and so four allowed transitions ($\Delta m_I = 1$), two of which involve H^A and two of which involve H^X (Figure 4.12). The arrangement where both nuclei are aligned "with" the applied field is the most stable and so lowest in energy, whereas that in which both nuclei are aligned "against" the magnetic field is the least stable and so highest in energy. X_1 is the energy required to excite H^X from being aligned "with" the applied field to "against" while H^A remains "with" the applied field, and X_2 is the energy required to excite H^X from being aligned "with" the applied field to "against" while H^A remains "against" the applied field. A_1 and A_2 are the corresponding transitions for the H^A nucleus.

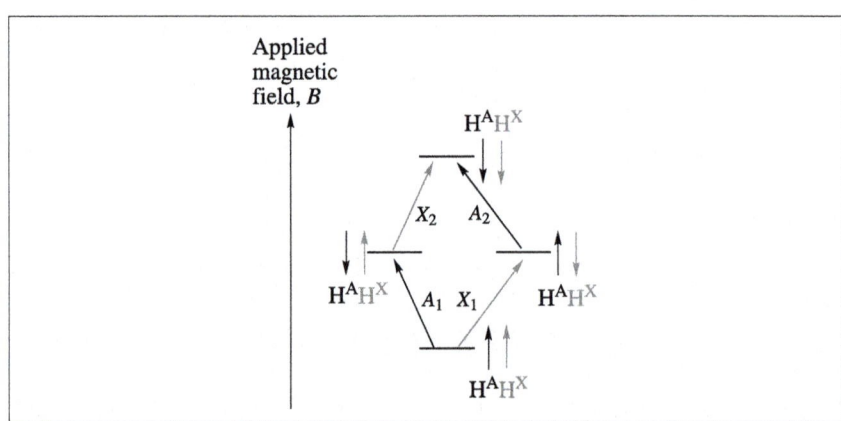

Figure 4.12   Energy levels of an AX spin system

We can see from the diagram that there is a difference in energy between the transitions X_1 and X_2, with X_1 being larger than X_2. This results in two resonances for H^X: X_1 at a higher frequency than if no H^A were present, and X_2 at a slightly lower frequency, *i.e.* a doublet centred on v_X. This situation would not have arisen without the presence of H^A and, consequently, there is an exactly equal and opposite effect on H^A by H^X, again leading to two resonances. Looking at the figure we can see that:

$$X_1 + A_2 = A_1 + X_2$$

so that:

$$X_1 - X_2 = A_1 - A_2$$

That is, the difference in energy between A_1 and A_2, and so the difference in the frequency of the transitions must be equal to the difference between X_1 and X_2 so that the splitting of the H^A and H^X signals is exactly equal (J_{AX}).

Predicting Splitting Patterns

Arrow Diagrams

We can predict spin–spin splitting patterns in either of two ways, and we will start by considering the effect on a given nucleus (nuclei) of the possible orientations in an applied magnetic field of the neighbouring protons. We will call this the "arrow diagram" method, in which nuclei aligned "*with*" the applied magnetic field are represented by an arrow *up*, and those "*against*" by an arrow *down*.

Consider nitroethane, $CH_3CH_2NO_2$, in which the methyl (CH_3) group has three chemically and magnetically equivalent protons, and the methylene (CH_2) group has two chemically and magnetically equivalent protons. The only coupling which will be observed, therefore, will be that between the three protons of the methyl group and the two protons of the adjacent methylene group. This coupling is often referred to as vicinal or three-bond (^3J) coupling and is the most frequently observed type of coupling in ^1H NMR spectra.

In the aliphatic region of ^1H NMR spectra, two-bond or geminal coupling (^2J) has the largest coupling constant (see later), but is less common than ^3J coupling, while four-bond coupling has the smallest coupling constant and is often non-existent. In general, then, the fewer the number of bonds between two nuclei which are coupled, the larger the coupling constant will be.

Coming back to nitroethane, in the "arrow diagram" method we need to consider what effect the methylene protons have on the methyl protons and *vice versa*. We can determine this by considering the possible orientations for the two methylene protons and the three methyl protons in an applied magnetic field.

Chemically and magnetically equivalent protons DO NOT couple to one another, so that the three CH_3 protons do not couple to one another and the two CH_2 protons do not couple to one another.

Geminal: combined in a pair. Vicinal: neighbouring (on adjacent atoms).

The CH_3 protons experience different magnetic environments, depending upon the orientation of the methylene protons (Box 4.7).

Box 4.7 Orientation of the Methylene Protons

(1) $\downarrow\downarrow$ $(\frac{1}{4})$

Both methylene protons are aligned "against" the applied field.

(2) $\uparrow\downarrow\ \downarrow\uparrow$ $(\frac{1}{2} = 2\times\frac{1}{4})$

One methylene proton "with" and one "against" the applied field. Note that this can occur in two different ways.

(3) $\uparrow\uparrow$ $(\frac{1}{4})$

Both methylene protons are aligned "with" the applied magnetic field.

The chemical shift of this triplet is quoted as the centre point of the triplet (the peak of relative intensity 2), which corresponds to the situation in which the effect of one of the methylene protons on the magnetic field of the methyl group is cancelled by the other (opposing spins), and so this peak has the same chemical shift as if there were no protons on the carbon adjacent to the methyl group.

Since there are a total of four possible orientations of these two protons – there are two ways of producing one proton "with" and one "against" – and, because in NMR the population difference between energy levels is incredibly small, the fraction of molecules in the sample with the methylene protons in each arrangement will be that given in parentheses (in Box 4.7) after the respective orientations. Thus, the CH_3 group adjacent to the methylene can exist in three magnetic environments and so gives three lines (a triplet) with relative probabilities, and so relative intensities, of the three lines of 1:2:1 (Figure 4.13a).

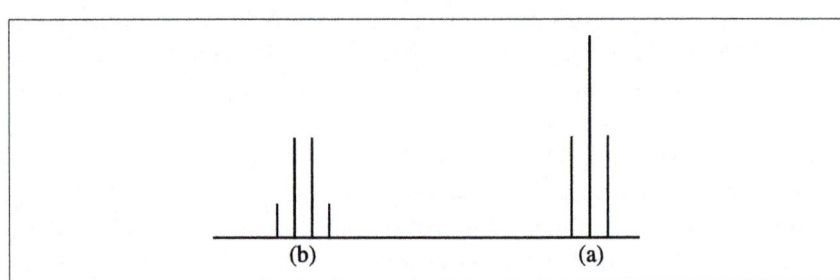

Figure 4.13 Predicted coupling pattern for the A_2X_3 spin system in nitroethane, $CH_3CH_2NO_2$

We can now look at the effect of the three methyl protons on the magnetic environment of the methylene group in the same way, but, as there are now three protons to be considered, there will be a greater number of possible orientations. Before looking at the pattern obtained (Box 4.8), you might like to try to work it out for yourself.

Box 4.8   Orientation of the Methyl Protons

(1) $\downarrow\downarrow\downarrow$ ($\frac{1}{8}$)

All 3 methyl protons aligned "against" the applied field.

(2) $\uparrow\downarrow\downarrow$ $\downarrow\downarrow\uparrow$ $\downarrow\uparrow\downarrow$ ($\frac{3}{8} = 3\times\frac{1}{8}$)

Two methyl protons "against" and one "with" the applied field. Note that this can occur in three different ways.

(3) $\uparrow\uparrow\downarrow$ $\uparrow\downarrow\uparrow$ $\downarrow\uparrow\uparrow$ ($\frac{3}{8} = 3\times\frac{1}{8}$)

Two methyl protons "with" and one "against" the applied field. Note that this can also occur in three different ways.

(4) $\uparrow\uparrow\uparrow$ ($\frac{1}{8}$)

All three methyl protons aligned "with" the applied magnetic field.

So, the methylene (CH_2) protons can exist in four magnetic environments (due to the different orientations of the three methyl protons) and so will give **four lines** (a **quartet**) in the ^1H NMR spectrum, of relative intensities 1:3:3:1 (Figure 4.13b).

Looking now at the actual ^1H NMR spectrum of nitroethane (Figure 4.14), we can see that there is a triplet of coupling constant (J) 7.4 Hz at δ 1.54 and a quartet of coupling constant 7.4 Hz at δ 4.38.

In this case you will notice that the centre point of the quartet does not correspond to a peak, so that the chemical shift is quoted as the midpoint of the four peaks.

When quoted as a list (*e.g.* in research papers), this spectrum would take the form: chemical shift (integral, splitting pattern, coupling constant), *i.e.*:
δ_H (300 MHz, CDCl$_3$) 1.54 (3H, t, $J = 7.4$ Hz), 4.38 (2H, q, $J = 7.4$ Hz).

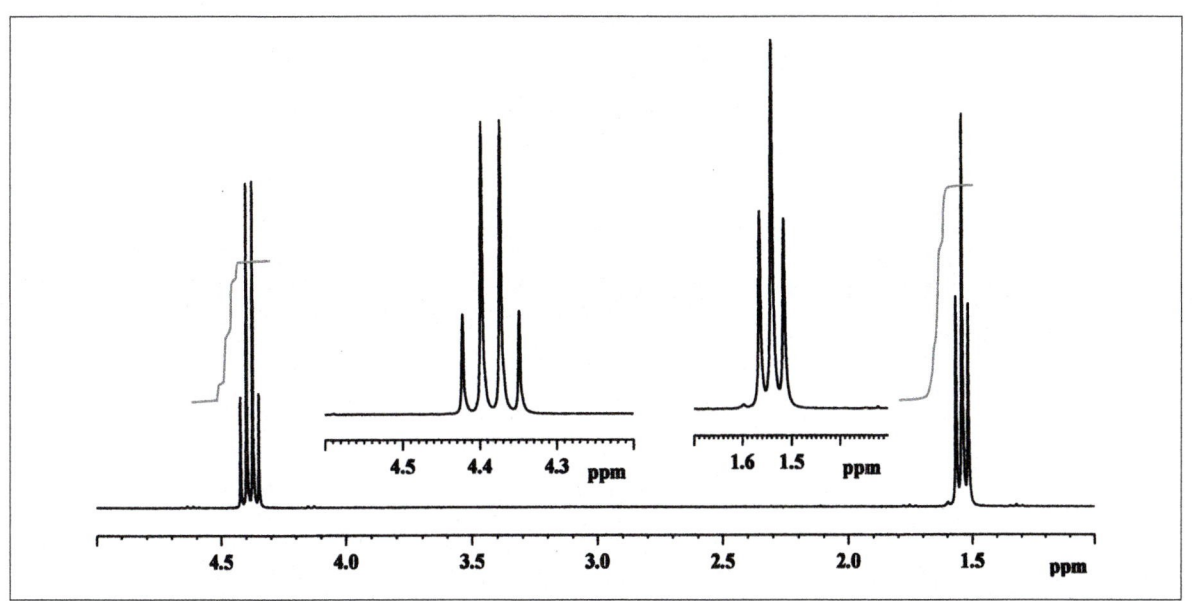

Figure 4.14   300 MHz ^1H NMR spectrum of nitroethane, $CH_3CH_2NO_2$, in $CDCl_3$

Notice that the integrals give us the ratio of the area under the peaks and not the absolute value, so that the spectrum of ethoxyethane ($CH_3CH_2OCH_2CH_3$), which would have exactly the same splitting pattern, would also have a triplet and a quartet in the ratio of 3H:2H.

Successive Splitting

A much more commonly used method for the prediction of splitting patterns is that of "successive splitting", since this method can be used (through the use of scale diagrams) for the prediction of the multiplicity (splitting pattern) of very complex spin systems.

In this method we take advantage of the fact that each *spin $\frac{1}{2}$ nucleus will split the peak due to its non-equivalent neighbour into a doublet of separation J (Hz)*. Interacting nuclei are represented by letters of the alphabet, with the gap between the letters in the alphabet indicating the separation between the nuclei in chemical shift. For example, in the description of the origin of spin–spin coupling, we used a two-nuclei system H^AH^X: this would be termed an AX spin system, in which the chemical shift separation for H^A and H^X is "large" enough to give first-order coupling (later we will look at the definition of first-order coupling, and what happens when two nuclei which are close in chemical shift, H^AH^B, couple). The spectrum of nitroethane (Figure 4.14) is representative of an A_2X_3 spin system.

AX Spin System

Two doublets with a coupling constant of J_{AX} (Figure 4.15).

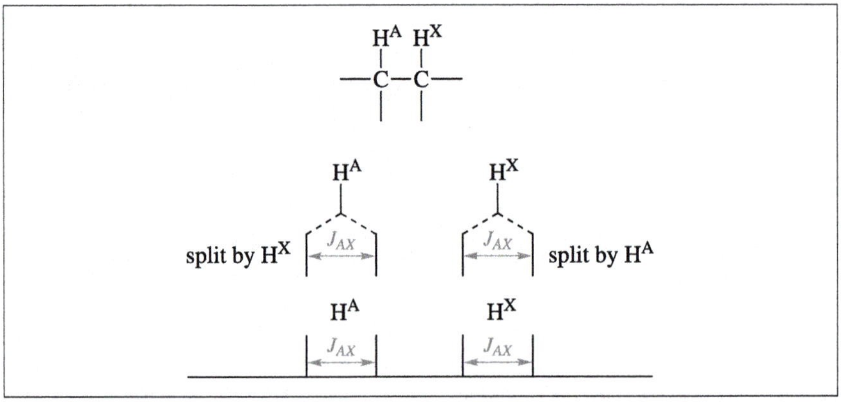

Figure 4.15 An AX spin system

AX$_2$ Spin System

A triplet (for H^A) of integral 1 and a doublet (for H^X) of integral 2 (Figure 4.16).

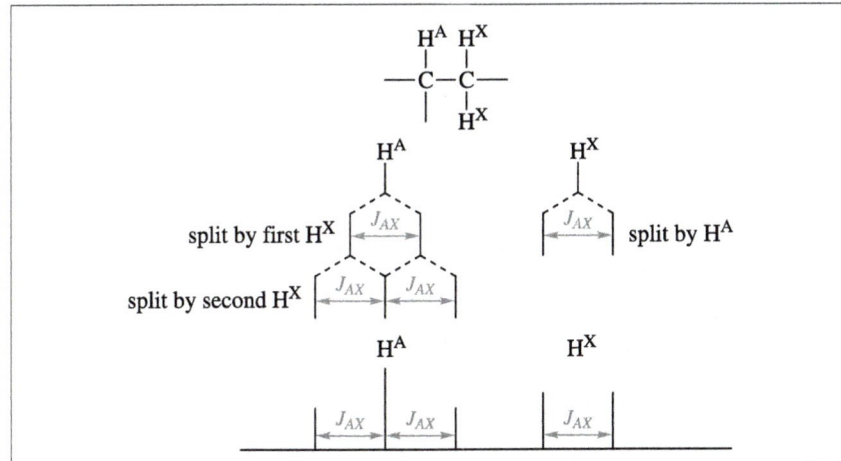

Figure 4.16 An AX$_2$ spin system

AX$_3$ Spin System

A quartet (for H^A) of integral 1 and a doublet (for H^X) of integral 3 (Figure 4.17).

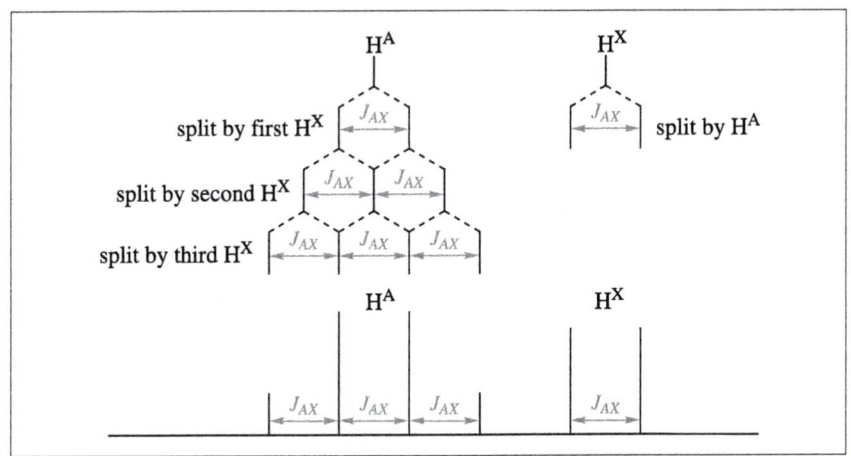

Figure 4.17 An AX$_3$ spin system

In general in ^{1}H NMR spectra, for n non-equivalent protons coupling to the proton in question, we will obtain $(n+1)$ lines, the relative intensities of which are given by Pascal's triangle (Box 4.9). Similar considerations do, of course, apply to other spin $\frac{1}{2}$ nuclei.

Box 4.9 Pascal's Triangle

We can see that the relative intensities of the outer lines in Pascal's triangle are always 1, and the intensities of the inner lines can be calculated by adding the two numbers diagonally to the left and right in the row above.

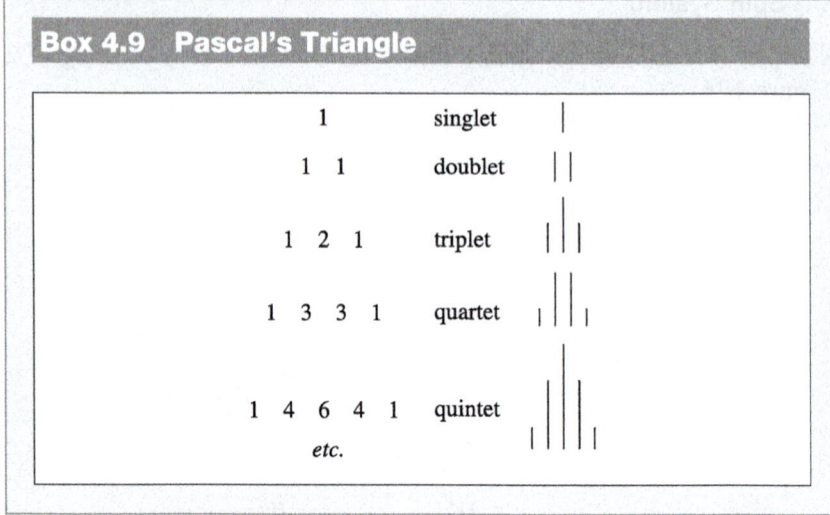

		1			singlet	
		1 1			doublet	
		1 2 1			triplet	
		1 3 3 1			quartet	
	1 4 6 4 1				quintet	
		etc.				

We normally do not see 1,4- (*para*) coupling as the *J* values are very small for five-bond coupling.

Let us now look at a more complex spin system, in which we have two different coupling constants to the same proton. In the trisubstituted aromatic 3-amino-4-methoxybenzoic acid shown in Figure 4.18, we have an AMX spin system and this leads to a 1,3- (or *meta*) coupled doublet for H^2 at δ 7.26, a 1,2- (or *ortho*) coupled doublet for H^5 at δ 6.85 and a doublet of doublets (both *ortho* and *meta* coupling) for H^6 at δ 7.21.

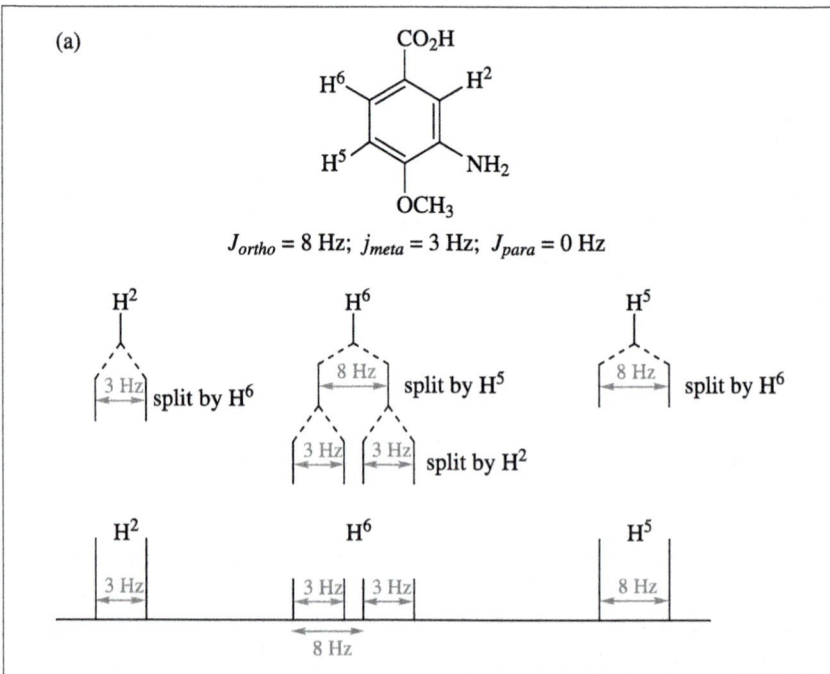

Figure 4.18 (a) Splitting pattern for 3-amino-4-methoxybenzoic acid

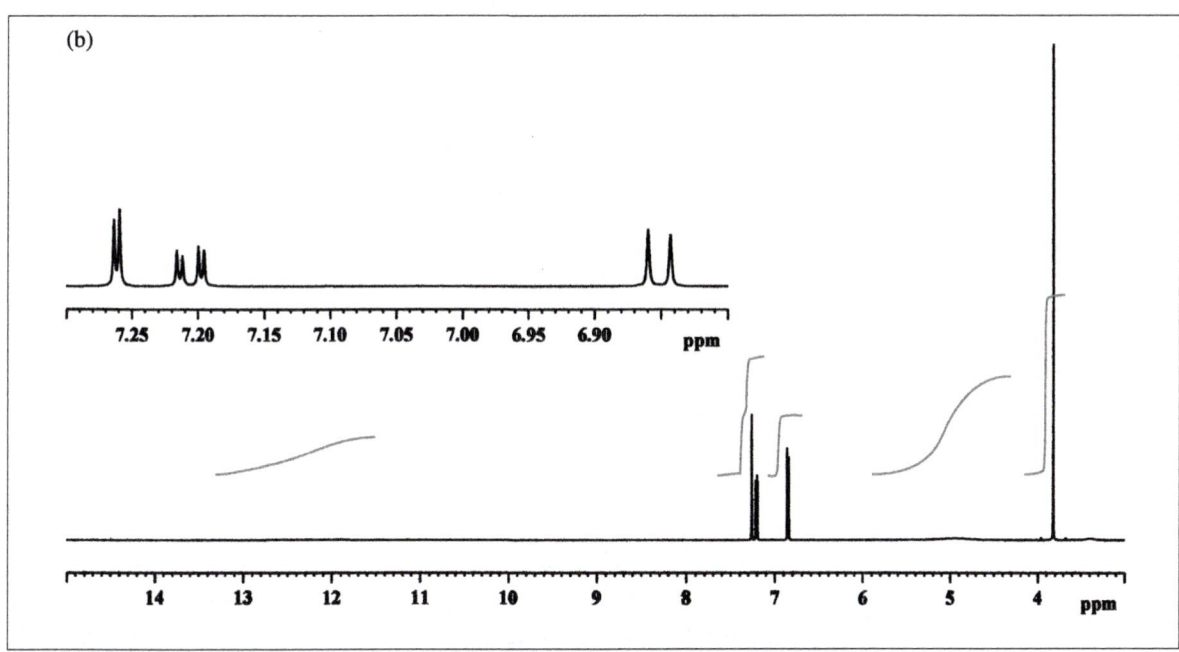

Figure 4.18   (b) Actual 500 MHz ^{1}H NMR spectrum of 3-amino-4-methoxybenzoic acid in DMSO-d_6

The simple patterns we have discussed so far give rise to "first-order" spectra which are very similar to the predictions we have made. First-order spectra are obtained when there is a reasonably large chemical shift difference between the coupling constants of the coupled nuclei, such that equation (4.4) is obeyed:

$$\delta_A - \delta_B(\text{in Hz}) \geq 10 \times J_{AB}(\text{in Hz}) \qquad (4.4)$$

where δ_A and δ_B are the chemical shifts of the two coupled nuclei, A and B, and J_{AB} is the coupling constant between these nuclei. From examining this equation, you may already have realized one advantage in running ^{1}H NMR spectra at as high a field as possible (*i.e.* at 500 MHz rather than at 300 MHz). Since the chemical shift of a signal is independent of the spectrometer frequency, running a spectrum at a higher magnetic field will mean that the difference in chemical shift between two signals will be the same in ppm, but the separation in hertz will be greater at the higher field. At higher field there is, therefore, more chance of obtaining first-order spectra.

Let us now look at what happens when the chemical shift difference between the coupled nuclei is small and the simple patterns become distorted, *i.e.* non-first-order spectra. The simplest example of this is an AB spin system. Remembering that an AX system will give a pair of doublets, we will now look at a system in which the chemical shift difference is relatively small, so that the spin system is now described as AB. The two doublets of the AX system now become distorted (with the amount of distortion from the AX spectrum increasing as the chemical shift difference decreases) and the line in each doublet which is closest to the other doublet (inner line) increases in size, while those furthest away (outer lines) decrease in size (Figure 4.19).

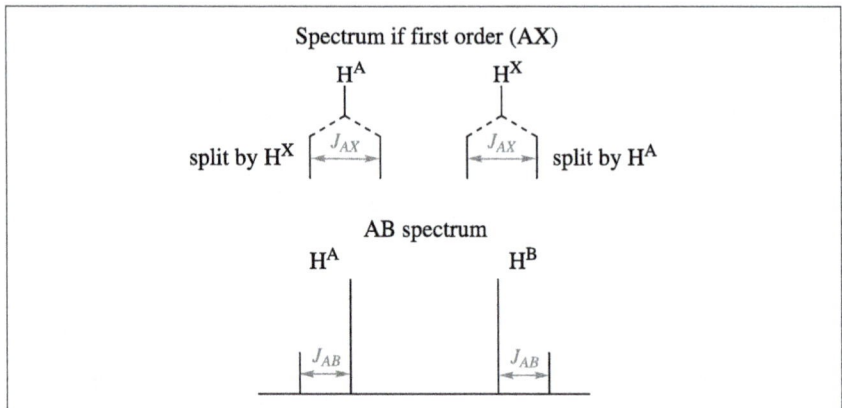

Figure 4.19 Non-first-order AB splitting pattern

This effect is often referred to as the "roof top" effect and is actually quite useful since the doublets slope towards the signal they are coupled to, often making the identification of coupled signals easier. Figure 4.20 shows a portion of the 500 MHz ^{1}H NMR spectrum of N-acetyl-L-phenylalanine, and we can clearly identify an ABX system due to the $-CH_2-CH-$ group. Normally the two protons of a methylene group (CH_2) would be chemically and magnetically equivalent and so would not couple to one another. However, if these protons are non-equivalent [which is often the case if they form part of a ring and/or are adjacent to a stereogenic (chiral) centre] then coupling between them occurs and, as they are likely to be close in chemical shift, since they are attached to the same carbon atom, an AB spectrum is obtained. As there is also an adjacent proton (the α-H of the amino acid), an ABX spectrum is obtained for N-acetyl-L-phenylalanine. The methylene protons (H^A and H^B) are adjacent to the chiral centre, so they are non-equivalent and couple to one another and to the proton attached to the stereogenic centre (α-proton), H^X. We have seen before, and shall see again in the following section, that the fewer the bonds between coupled nuclei, the greater the coupling constant, so the geminal (^2J) coupling of the methylene protons to each other will be greater than the coupling of either H^A or H^B to H^X. In addition, H^A and H^B frequently have non-equivalent vicinal (^3J) coupling constants to H^X. In the 500 MHz spectrum of N-acetyl-L-phenylalanine (Figure 4.20), we can see the pair of doublets of doublets (one for H^A and one for H^B) at δ 2.85 and 3.05. Closer examination of this spectrum shows that the doublet of doublets at δ 2.85 has coupling constants of 13.8 and 9.6 Hz, while that at δ 3.05 has coupling constants of 13.8 and 4.9 Hz. The larger coupling constant (13.8 Hz) is obviously the geminal coupling constant, J_{AB}, while the two smaller coupling constants (9.6 and 4.9 Hz) are the vicinal coupling constants, J_{AX} and J_{BX}. The signal for H^X will be complicated as this proton couples to both H^A and H^B (with J values of 9.6 and 4.9 Hz), and also to the NH (coupling constant, $J = 7.7$ Hz) and can be seen at δ 4.40.

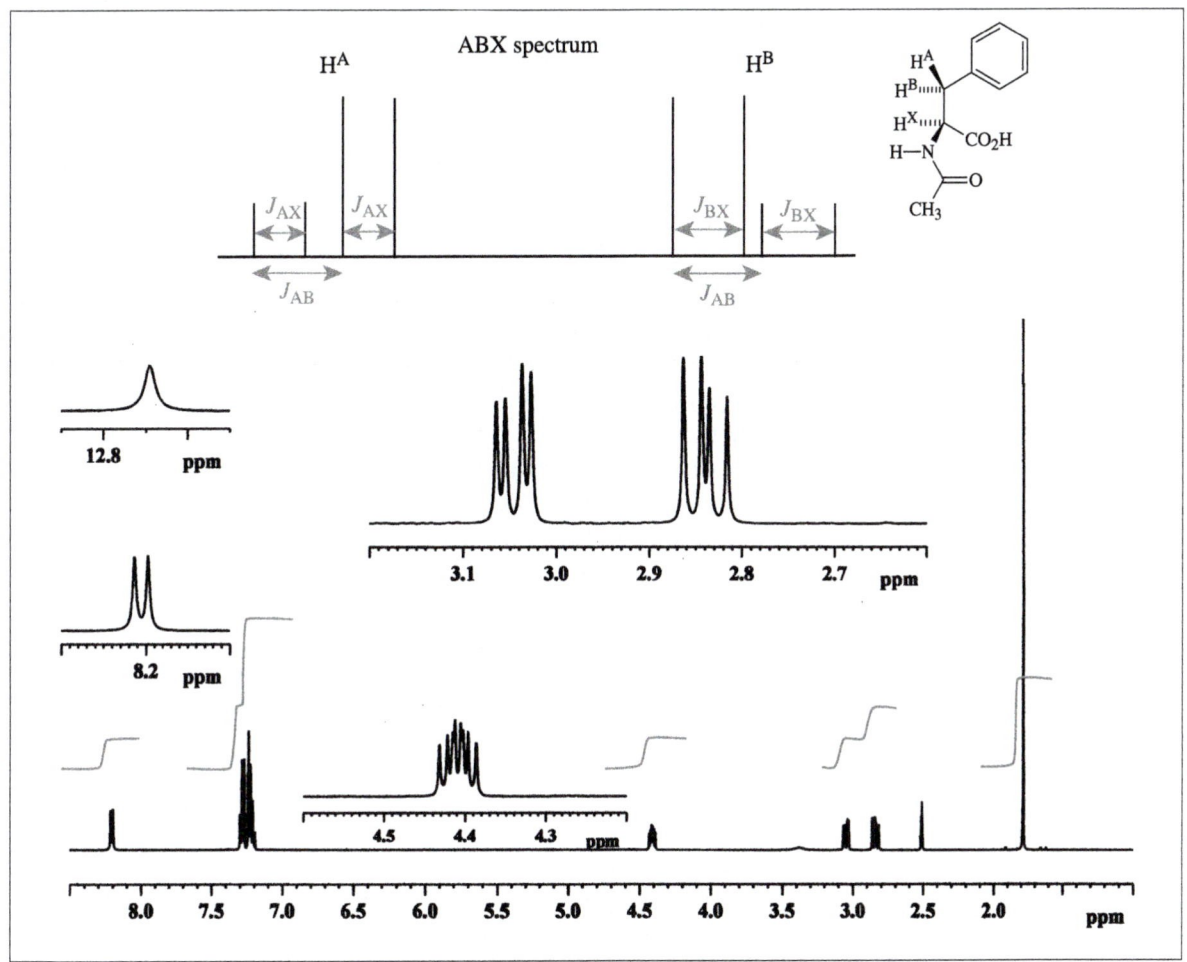

Figure 4.20　500 MHz ^{1}H NMR spectrum of N-acetyl-L-phenylalanine in DMSO-d_6

The spectrum of N-acetyl-L-phenylalanine also introduces us to another important consideration in ^{1}H NMR spectra: the effect of **stereoisomerism** on the ^{1}H NMR spectrum. Each of a pair of enantiomers has the same physical properties (apart from their ability to rotate the plane of plane polarized light) and so will have identical ^{1}H NMR spectra. The spectra for N-acetyl-L-phenylalanine and its enantiomer, N-acetyl-D-phenylalanine, will therefore be identical in every respect. Diastereoisomers, on the other hand (no pun intended), have different physical properties and so give different ^{1}H NMR spectra.

Enantiomers are non-superimposable mirror images whereas diastereoisomers are stereoisomers which are not mirror images.

In an extension of this effect, enantiotopic (prochiral) groups [groups which, if different, would produce a chiral (stereogenic) centre] do not lead to different ^{1}H NMR spectra, but diastereotopic groups (groups which, if different, would produce diastereoisomers) do. To illustrate this point, we can use an imaginary substitution test and imagine replacing one of the hydrogens of (chloromethyl)benzene or dichloroacetaldehyde diethyl acetal (1,1-dichloro-2,2-diethoxyethane) by deuterium (Box 4.10). We can see that the two hydrogens of the methylene group in (chloromethyl)benzene are prochiral but substitution would give rise to enantiomers, so that in ^{1}H NMR terms these protons are equivalent. For dichloroacetaldehyde diethyl acetal, however, we can see that substitution of either of the two hydrogens of the ester methylene group gives a stereogenic centre at this carbon *and* at the carbon which has the two ethoxy groups attached. In this case, the substitution test would lead to diastereoisomers, so these methylene protons are non-equivalent and couple to one another, as well as to the adjacent methyl protons, to give two overlapping sets of eight lines at δ 3.75 (Figure 4.21). This imaginary test is often very useful in predicting whether the two protons of a methylene group couple to one another, but it is not infallible, and often two diastereotopic protons do not couple to one another (especially if they are remote from the other stereogenic centre in the molecule).

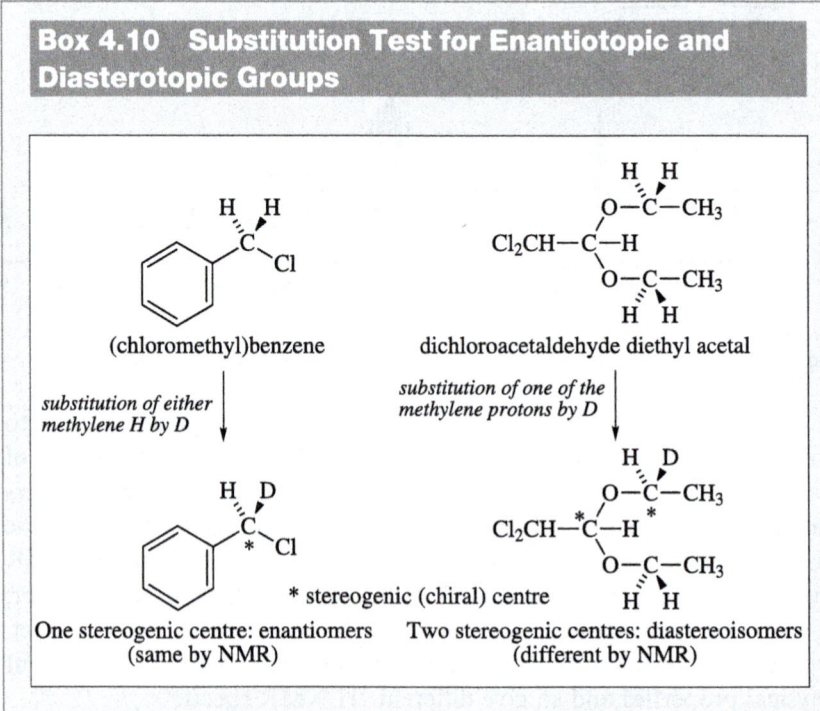

Box 4.10   Substitution Test for Enantiotopic and Diasterotopic Groups

(chloromethyl)benzene

dichloroacetaldehyde diethyl acetal

substitution of either methylene H by D

substitution of one of the methylene protons by D

* stereogenic (chiral) centre

One stereogenic centre: enantiomers (same by NMR)

Two stereogenic centres: diastereoisomers (different by NMR)

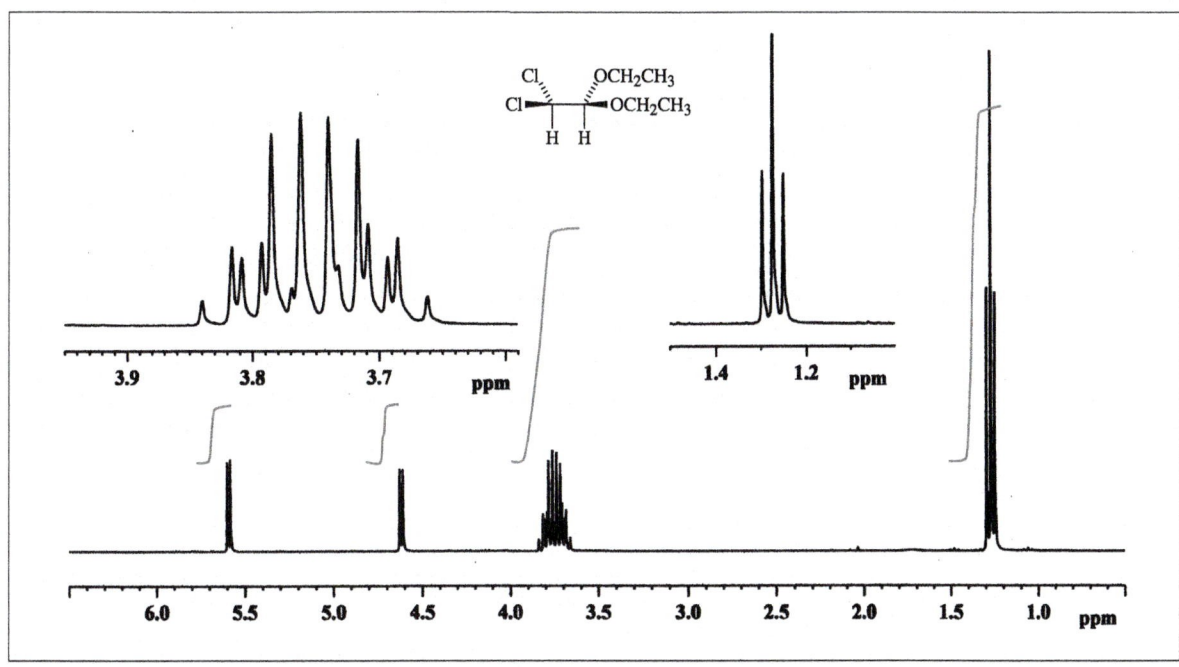

Figure 4.21 300 MHz ^{1}H NMR spectrum of dichloroacetaldehyde diethyl acetal in CDCl$_3$

How do we know if a peak is a doublet of triplets (dt) or a triplet of doublets (td)? The multiplet with the largest coupling constant is quoted first, *i.e.* a doublet of triplets (dt) has the larger *J* value for the doublet (see Box 4.11).

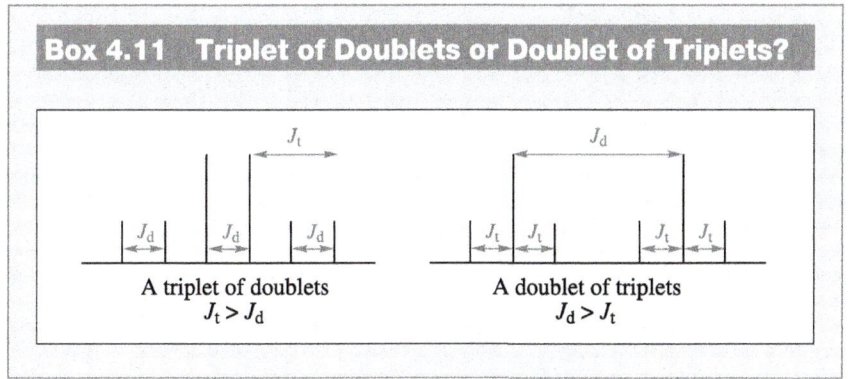

Box 4.11   Triplet of Doublets or Doublet of Triplets?

A triplet of doublets
$J_t > J_d$

A doublet of triplets
$J_d > J_t$

The Magnitude of Spin–Spin Coupling

As has been stated a number of times in previous sections, the size of the coupling constant, *J*, is dependent upon the number of bonds between the coupling nuclei and the molecular environment of the nuclei. It is generally true that the fewer the bonds between nuclei which are coupled, the greater the coupling constant. For example, as we shall see in the following sections, one-bond ^{13}C–^{1}H coupling constants typically lie in the range 120–250 Hz, and, as we have seen already, aliphatic geminal (two-bond, ^{2}J) ^{1}H–^{1}H coupling constants often lie in the range 10–20 Hz and aliphatic vicinal (three-bond, ^{3}J) ^{1}H–^{1}H coupling constants in the range 5–8 Hz. However, the difference between the magnitude of ^{2}J and ^{3}J coupling constants is not clear-cut. There are a number of ^{3}J coupling constants which lie in the 10–20 Hz range, most notably the coupling constant for *trans* alkene protons (12–18 Hz), and a number of ^{2}J coupling constants which lie in the 0–8 Hz range, most notably that between the two protons of a terminal alkene group (0–3 Hz). Longer range (^{4}J and ^{5}J) couplings can be observed, especially in aromatic or alkenic molecules, but these are

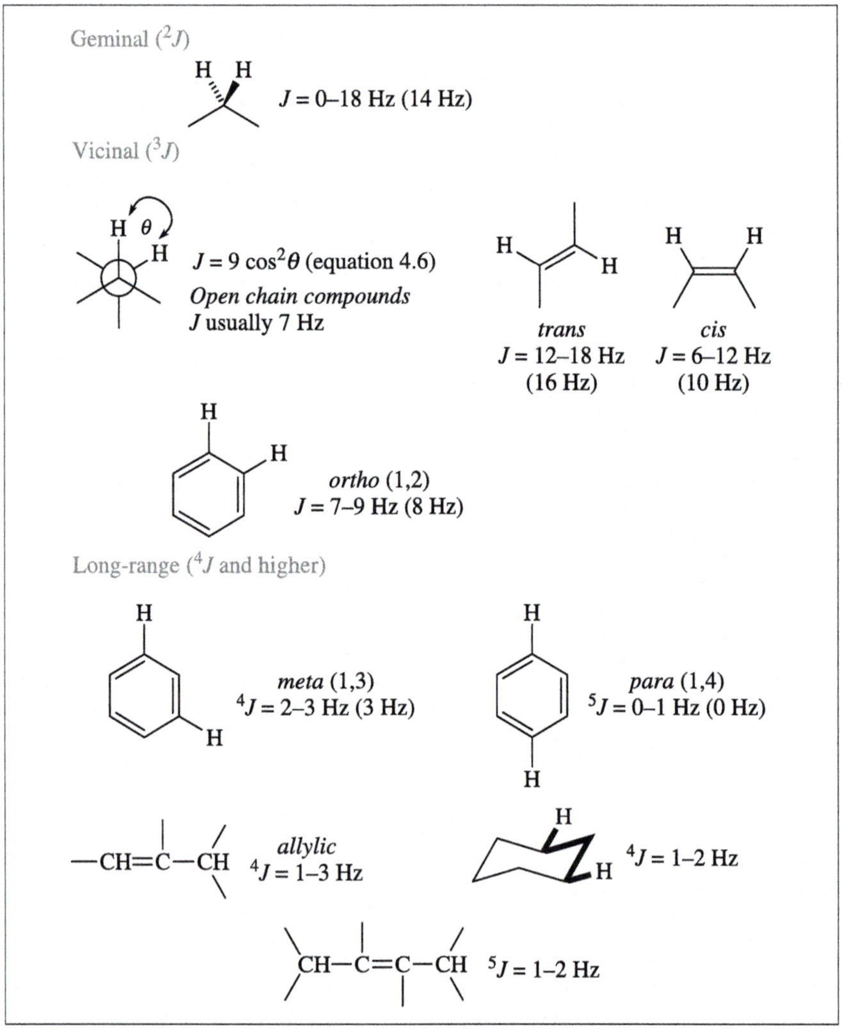

Figure 4.22 Examples of ^{1}H–^{1}H coupling constants (usual values given in brackets)

less common. Figure 4.22 contains a list of some representative coupling constants and extensive lists are available elsewhere. We will now briefly consider those factors which have the greatest influence upon the magnitude of the coupling constant, but we should always remember that, for all coupling, the shorter (and so stronger) the bonds between the nuclei which are coupling, the larger the coupling constant will be. Since most structural information can be obtained from vicinal coupling constants, we will concentrate initially upon the factors which influence the magnitude of this three-bond coupling.

Vicinal coupling constants can give a great deal of information about the conformation adopted by a molecule. Such conformational analysis is particularly useful in cyclic systems and in peptides/proteins, which have some conformations which are more energetically favoured than others.

Conformations can be interconverted by rotation about a single bond, while configurations are not interconvertible by rotation.

Remembering that coupling involves the interaction of the nuclear spins of the coupling nuclei *via* the bonding electrons of the bonding system between them, it becomes apparent that the coupling constant is dependent upon orbital overlap, and so on the dihedral angle, θ, between the planes containing the nuclei. The relationship between the vicinal coupling constant between nuclei H^A and H^X, J_{AX}, and the dihedral angle, θ, is given by the Karplus equation, equation (4.5):

$$^3J_{AX} = J^S \cos^2\theta - 0.28 \text{ (Hz)} \qquad (4.5)$$

where J^S is a constant dependent upon the other substituents present on the carbon atoms. A very useful *approximation*, however, is equation (4.6):

$$^3J_{AX} \approx 9\cos^2\theta \text{ (Hz)} \qquad (4.6)$$

i.e. $J^S \approx 9$, which gives the plot shown in Figure 4.23.

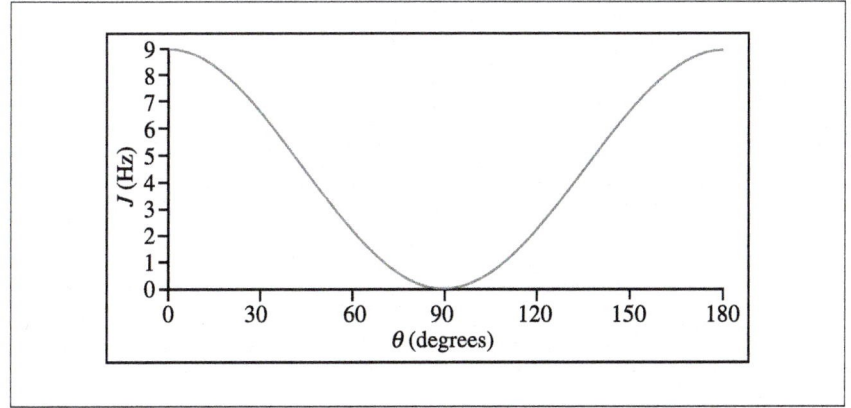

Figure 4.23 Plot of dihedral angle (θ) against coupling constant (J)

Note that the value of J^S is actually slightly greater than 9 Hz at $\theta = 180°$ (where the C–H orbitals have the greatest degree of overlap) and slightly lower than 9 Hz at $\theta = 0°$. As $\cos 90° = 0$, the coupling constant due to a dihedral angle of 90° is 0 (where there is the least amount of overlap between orbitals). As mentioned previously, the Karplus equation is particularly useful in more rigid systems (although the accuracy of the predicted values is dependent upon the substituents present), and a good example of its use in cyclic systems is in the analysis of the coupling constant of the anomeric hydrogens in methyl α- and β-D-glucopyranosides. In methyl α-D-glucopyranoside (Figure 4.24), the coupling constant – due to coupling to the hydrogen on C2 – for the anomeric hydrogen (H^1), δ 4.52, is 3.6 Hz, which is larger than the predicted value of 2.25 Hz for a dihedral angle (θ) of 60°. In methyl β-D-glucopyranoside, the coupling constant for the anomeric hydrogen is 7.8 Hz, which is smaller than the predicted value of 9 Hz for a dihedral angle of 180°.

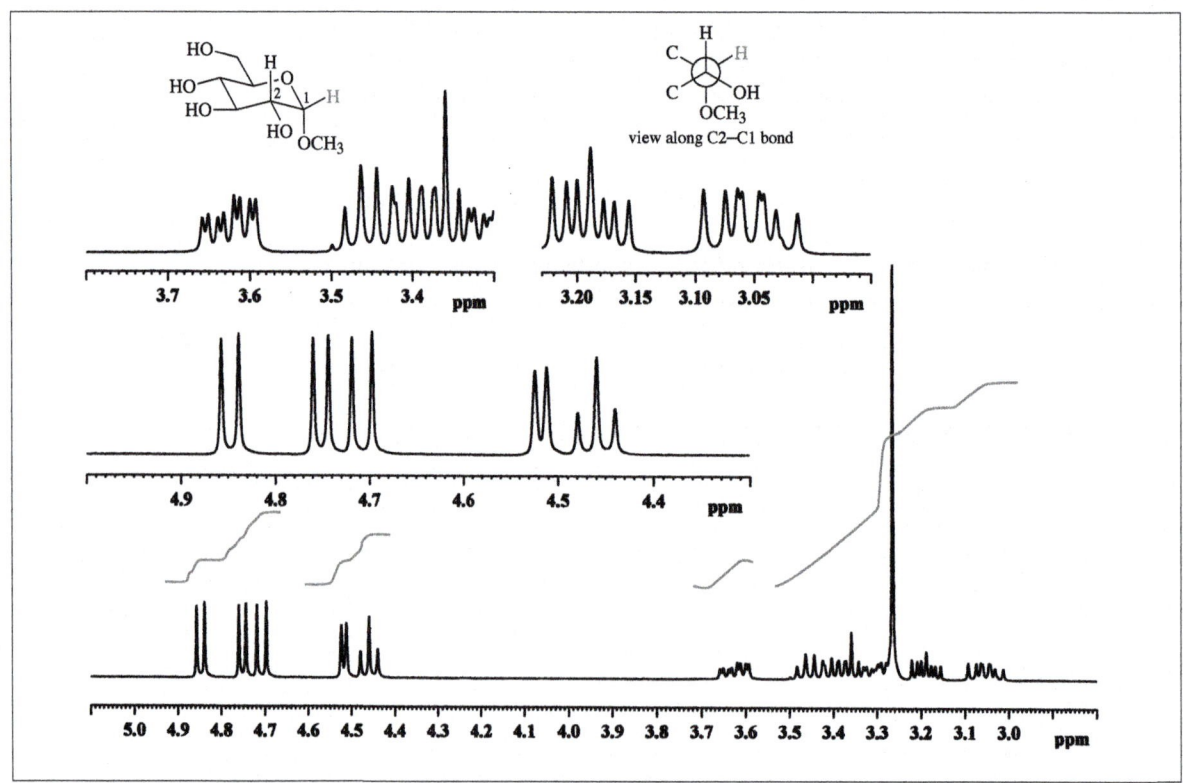

Figure 4.24   300 MHz ^{1}H NMR spectrum of methyl α-glucopyranoside in DMSO-d_6

Conformational analysis is also possible for the secondary structure present in peptides and proteins. For example, in helices the dihedral angle H–N–C$_\alpha$–H is ~120° while in β-sheets it is ~180°, so that these structural features can be identified by the coupling constant of the α-H of $J < 6$ Hz for helices and $J > 7$ Hz for a β-sheet.

It is difficult to find simple examples to illustrate the effect of angle strain on the vicinal coupling constant, but a reduction in the angle strain results in an increase in the vicinal coupling constant.

The presence of electronegative elements directly attached to the same carbon atom as one of the vicinally coupled protons decreases the magnitude of the coupling constant, while the presence of electropositive elements increases it. This effect is small in chains (which are capable of relatively free rotation) but more pronounced in rigid systems such as alkenes.

Geminal coupling constants are increased (made more negative) by:

- The introduction of an electronegative substituent on an adjacent carbon atom.
- The introduction of an electropositive substituent on the same carbon atom as the geminally coupled hydrogens.
- Having an adjacent π-bond, so that the C–H orbital can overlap with the π-orbitals of the π-bond (this effect is more pronounced for carbonyls than for carbon–carbon double bonds).

They are decreased by:

- An increase in the H–C–H bond angle (*i.e.* an increase in the %s character of the carbon atom).
- The introduction of an electronegative substituent on the same carbon atom as the geminally coupled hydrogens.
- The introduction of an electropositive substituent on an adjacent carbon atom.

Long-range couplings are generally limited to ^4J or ^5J, and then only in exceptional circumstances, *e.g.* the short bonds in aromatic rings give rise to *meta* (^4J) and occasionally *para* (^5J) coupling between protons, and the even shorter bonds present in alkenes and alkynes can give rise to even longer range couplings (up to ^7J). Finally, in saturated systems, a particularly favourable arrangement for long-range coupling is when four bonds adopt a W arrangement (highlighted in bold in Figure 4.22).

Angle strain: the strain due to the bond angles in a ring system being non-ideal, *e.g.* cyclobutane bond angles are close to 90° but the "ideal" sp³ (tetrahedral) angle is 109.5°.

In the mathematical treatment of coupling, vicinal (^3J) and ^1J (^{13}C–^1H) coupling constants are positive in sign and geminal (^2J) coupling constants are negative, but this has no effect upon the appearance of the spectrum.

4.5 Decoupled, NOE (Double Resonance) and COSY Spectra

4.5.1 Decoupled Spectra

If it is not obvious which protons in a molecule are coupled to which, we can determine coupled signals through the use of decoupling experiments (Figure 4.25), in which we irradiate one nucleus while detecting all the other nuclei in the molecule, *i.e.*"double resonance" using two pulses of radiation. In this experiment, the signal due to any nucleus coupled to that being irradiated collapses, and the peak obtained is that which we would see if the nucleus being irradiated were not present. By irradiating at the resonance frequency of a specific nucleus we are continually exciting this nucleus to its higher energy state, which is also continually relaxing back to its lower energy state. This is happening so fast on the NMR timescale (faster than once per millisecond) that any nuclei coupled to the irradiated nucleus "see" only the average of the two spin states (α and β), and any coupling to it is removed. With the ready availability of COSY spectra (see the next section), this technique is not employed as often as it once was. In Figure 4.25b, the triplet at δ 4.32 for the methylene in bold in Figure 4.25a was irradiated, causing the methylene protons which were coupled to it (red) to lose all coupling to these protons. The methylene signal at δ 1.74 therefore collapses from a quintet (five lines) to a triplet (since it is now coupled only to the methylene, CH_2, protons) and the signal being irradiated disappears, with the rest of the spectrum being unchanged.

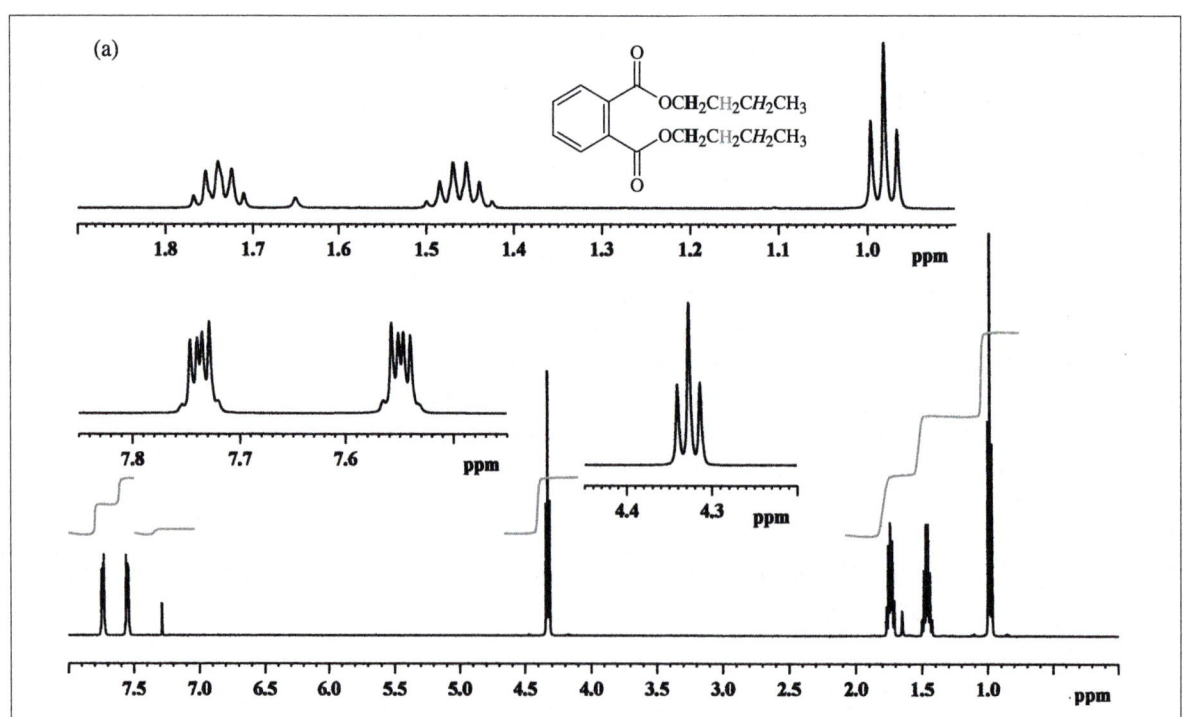

Figure 4.25   (a) 500 MHz ^{1}H NMR spectrum of dibutyl phthalate in CDCl₃

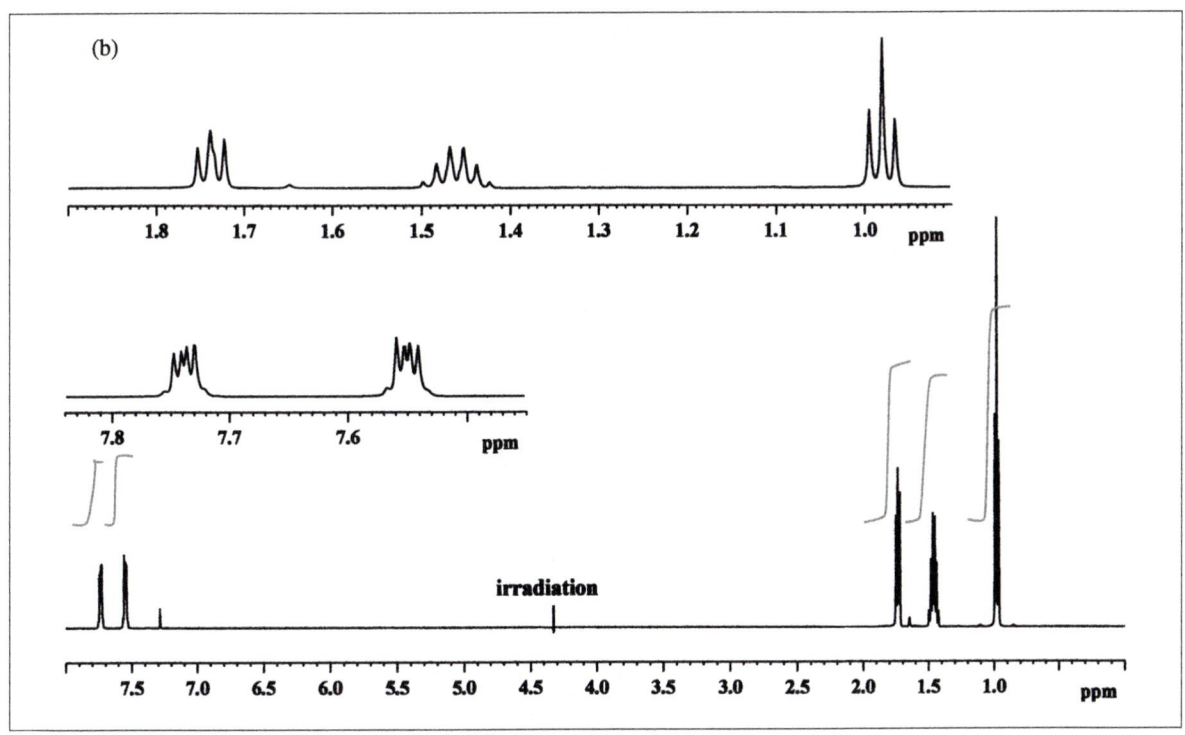

Figure 4.25 (b) Decoupled (double resonance) 500 MHz ^{1}H NMR spectrum of dibutyl phthalate in CDCl$_3$, with irradiation at δ 4.32

4.5.2 COSY Spectra

The acronym COSY stands for COrrelated SpectroscopY and this technique is widely used to determine all of the coupling interactions in a single experiment. This proves to be more efficient than the decoupling experiment in which each signal is irradiated in turn to determine its coupling partners. COSY involves a multiple pulse sequence (which we do not need to know anything about in order to use the technique) and is an example of two-dimensional (2D) spectroscopy.

The HH COSY experiment is quick to perform, relatively easy to interpret and allows us to correlate the ^{1}H shifts of all the coupling partners in the molecule. The "normal" ^{1}H spectrum is plotted on both the x- and y-axes, and a projection also appears along the diagonal (with the peak heights represented by contours). The key signals in this spectrum are the correlation (or cross) peaks which appear off the diagonal. The COSY spectrum is symmetrical about the diagonal axis, and the cross-peaks therefore all appear on either side of the diagonal axis and are symmetrical about it. The symmetry of the spectrum makes identification of the cross-peaks for a given signal relatively simple, since the two coupled peaks (on the diagonal) and the cross-peaks (off the diagonal) form the four corners of a square. In order to determine the coupling partner for any peak, we need only go directly vertically (or horizontally) from a given peak until we come to a cross-peak. Going horizontally (or vertically) from this peak to the diagonal takes us to the peak for the coupling partner (see Figure 4.26).

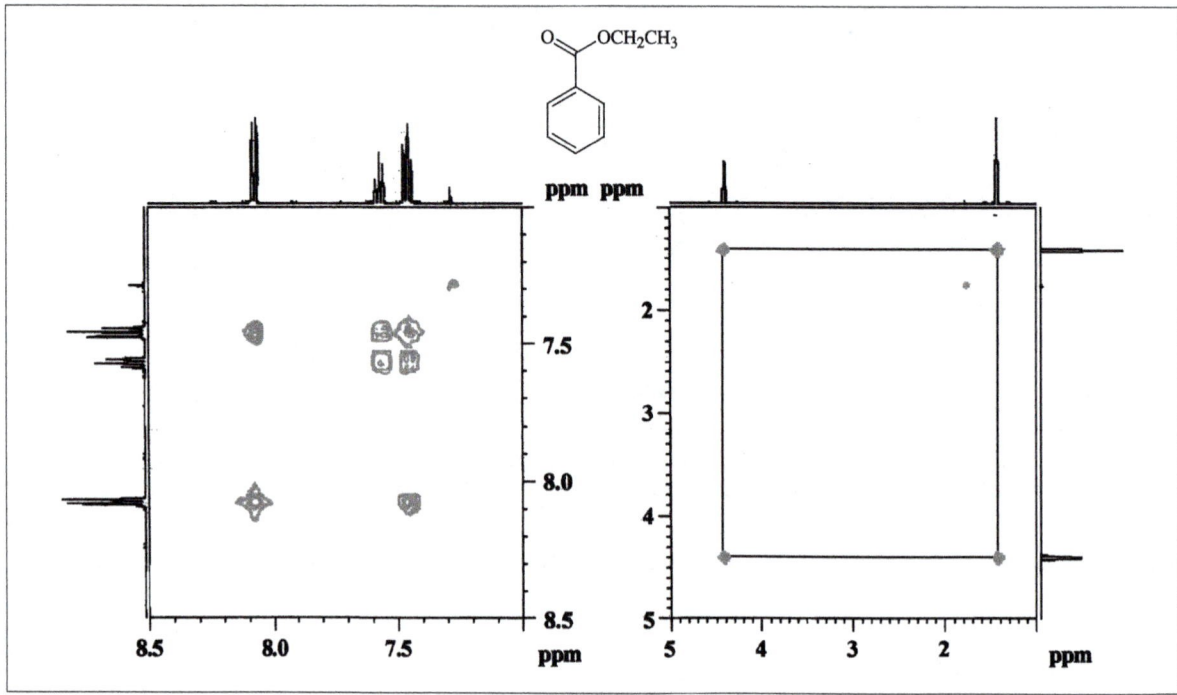

Figure 4.26   500 MHz HH COSY NMR spectrum of ethyl benzoate in CDCl₃

You should try to assign the aromatic region of the ¹H NMR spectrum of ethyl benzoate using the expansion of the δ 6.5–8.5 region of the COSY spectrum (Figure 4.26).

As all the coupling relationships in the molecule are given in the COSY spectrum, it is normal to be able to deduce all of the structural fragments in a molecule from the HH COSY spectrum.

4.5.3 NOE Spectra

If we irradiate one nucleus, while detecting another nucleus which is close to it in space (but not necessarily bonded to it), *i.e.* double resonance, then the signal for the detected nucleus is enhanced and the signal-to-noise ratio improved. We can achieve this by irradiating a given nucleus with a low-level radiofrequency signal and monitoring what is happening in the rest of the spectrum. Because the NMR relaxation processes (those processes which help the system to return to equilibrium where there is an excess of nuclei in the lower energy level) are governed by through-space interactions, enhancement of the signals for *nearby* nuclei (the nuclear Overhauser effect, NOE) is observed. We do not need to understand the theory behind this effect, which is very useful in obtaining distance information in a molecule, *e.g.* which protons are close in space to each other in a molecule. One of the main uses for this effect is to study the conformation of proteins, but we can also take advantage of this effect in decoupled ^{13}C spectra, and it will be discussed again in this context later.

As the enhancements we see due to ^1H–^1H NOE are small (typically 7–10%), it is extremely difficult to see any effect by simple comparison of two spectra, with and without the NOE effect. The easiest way to analyse these type of data is by difference spectroscopy, with a control spectrum acquired in which the NOE effect has deliberately not been switched on and a second spectrum acquired where the NOE has been allowed to build up. Upon subtraction of the control spectrum from the spectrum with the NOE present, we obtain an NOE difference spectrum in which the signals with enhancements show up as small positive peaks, the signal which was irradiated shows up as a large negative peak, and all other (unaffected) peaks are absent (Figure 4.27). These spectra are usually quoted with the % enhancement of the signals. To measure this we need to take into account the integral for the peak irradiated (the one signal which is negative on the difference spectrum). For example, in Figure 4.27b, the signal at δ 3.60 is for 3H (OCH₃), so 1H will be equivalent to $^1/_3$ of this value.

The % enhancement is found by dividing the height of the integral for the peak of interest by the value for 1H, and multiplying by 100. For the peak at δ 4.42, this gives an enhancement of 5.9% (the maximum value is 50% but usual values are between 1 and 20%).

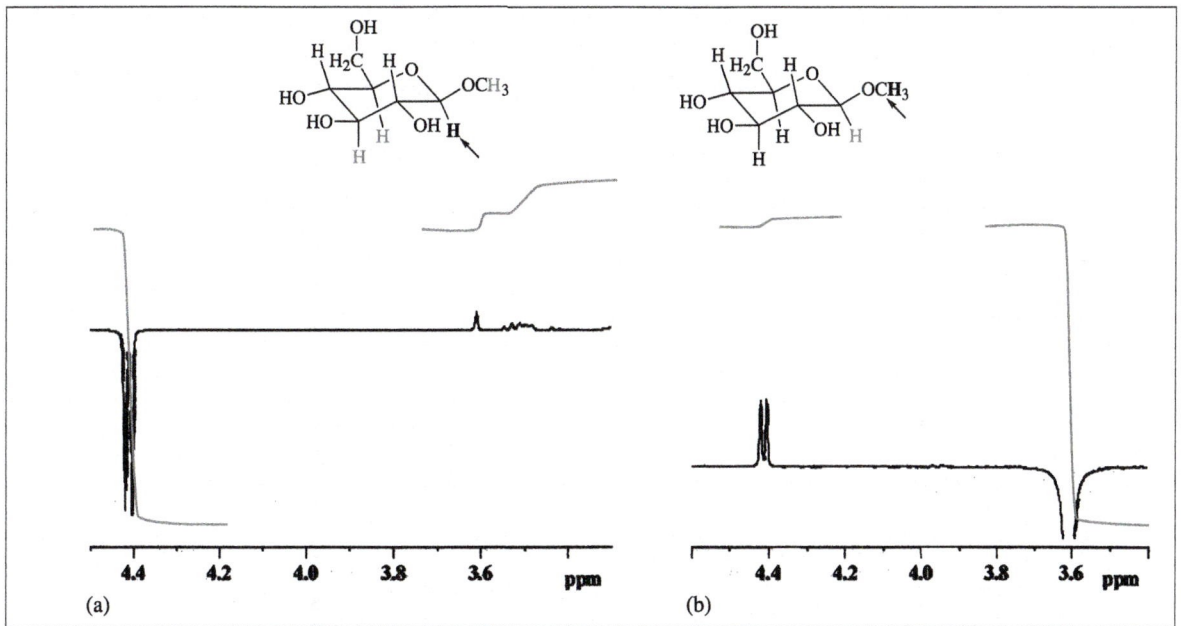

Figure 4.27   500 MHz NOE spectra of methyl β-glucopyranoside in DMSO-d_6: (a) irradiation of the anomeric hydrogen (δ 4.42), giving enhancement of the methoxy signal and ring hydrogens (*red*); (b) irradiation of the methoxy signal (δ 3.60), giving enhancement of the anomeric position (*red*)

Once again, there is a 2D version of this experiment which gives all of the NOE effects in the molecule in a single spectrum (NOESY). This experiment works best for large molecules ($m/z > 1000$), but can

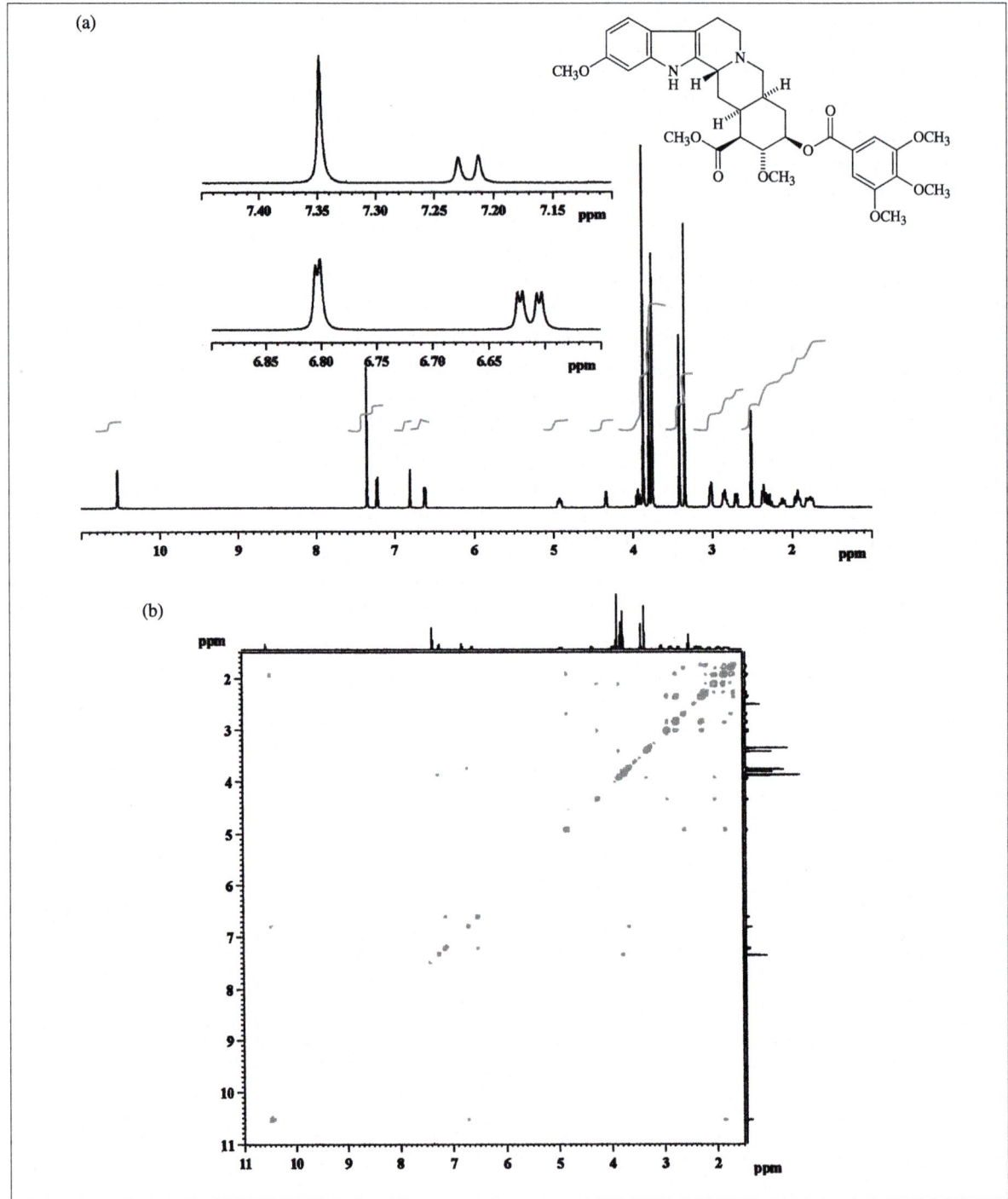

Figure 4.28   (a) 500 MHz ^{1}H NMR spectrum of reserpine in DMSO-d_6; (b) 500 MHz NOESY NMR spectrum of reserpine in DMSO-d_6

sometimes be used for relatively small molecules (Figure 4.28). Cross-peaks in the NOESY spectrum indicate NOE between the two peaks to which they give correlations, and the hydrogens giving rise to these peaks must therefore be close in space.

4.6 ^{13}C NMR Spectroscopy

Carbon-12 has an abundance of 98.9% but a nuclear spin (I) of 0 and is not detectable by NMR. Fortunately, ^{13}C, with a natural abundance of 1.1%, has a nuclear spin of $\frac{1}{2}$ and so gives rise to NMR signals, and ^{13}C NMR spectroscopy has become an important tool in structure confirmation and elucidation.

Since only 1.1% of the carbon atoms in a molecule are detectable (by ^{13}C NMR), the data are more difficult to acquire than the corresponding ^{1}H data. However, even a few milligrams of sample have billions of molecules present, so that there is a very high probability of each carbon atom being ^{13}C in some of the molecules.

Problems of sensitivity are largely overcome by using a greater number of scans or more concentrated samples. Typically, we can obtain a ^{1}H NMR spectrum using 4–16 scans on 10–20 mg of sample, whereas the corresponding ^{13}C data will require 256–2048 scans to obtain a reasonable spectrum.

A conventional ^{13}C NMR spectrum is referred to as being proton decoupled (Figure 4.29) and consists of a plot of chemical shift against intensity, with one peak for each carbon atom in the molecule (the three peaks at δ 79.8 are for the solvent, CDCl$_3$). As with proton NMR spectroscopy, chemically and magnetically equivalent atoms appear together on the spectrum; however, unlike a ^{1}H NMR spectrum, the area under each peak is not always proportional to the number of ^{13}C nuclei giving rise to the peak. In particular, quaternary carbons (which have no attached protons) have no nearby nuclei with spin to which they can transfer their energy, so they do not have sufficient time to "relax" back to equilibrium (with the excess population in the lower energy state) before the next pulse is applied. As a result, their peak intensities are normally reduced compared to other groups, *e.g.* methyl (CH$_3$). It is for this reason that we do not integrate the ^{13}C spectrum, as the peak areas are not necessarily related to the number of carbon atoms in each molecular environment.

Look at the structure of methyl α-glucopyranoside (Figure 4.29) and satisfy yourself that it has seven non-equivalent carbon atoms. You might also try to assign the two extreme signals.

For a given magnet the ^{13}C frequency is about one quarter the value of the ^{1}H frequency, *e.g.* on a 300 MHz instrument the ^{1}H frequency will be 300 MHz and the ^{13}C frequency will be 75 MHz. In addition, the resonance frequency of ^{13}C nuclei is vastly different from that of ^{1}H nuclei, so that no part of the ^{1}H and ^{13}C shift ranges overlap.

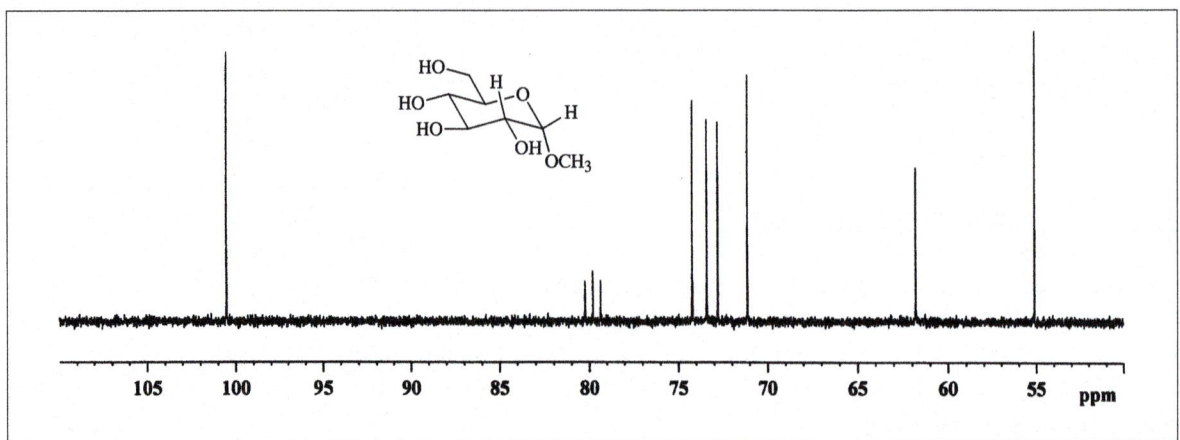

Figure 4.29   75 MHz ^{13}C NMR spectrum of methyl α-glucopyranoside in a mixture of CDCl$_3$ and DMSO-d_6; δ$_C$ CDCl$_3$ 79.8 (three peaks), δ$_C$ DMSO-d_6 40.5 (not shown)

4.6.1 ^{13}C Chemical Shifts

The chemical shifts of carbon atoms usually lie between δ 0–220 (with TMS again being used as the reference value of δ 0) (Boxes 4.12–4.14),

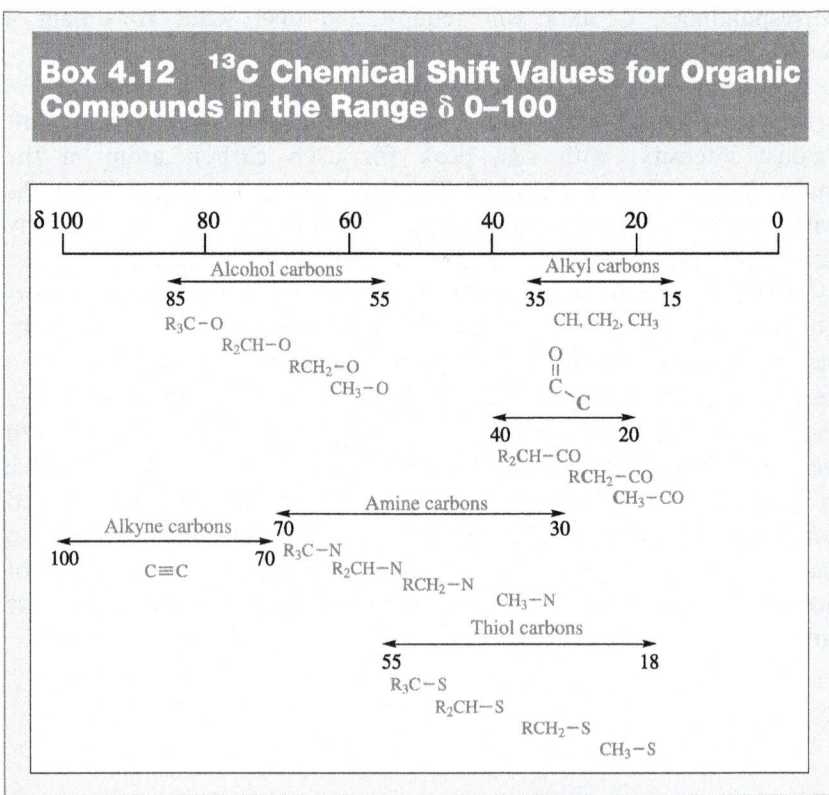

Box 4.12 ^{13}C Chemical Shift Values for Organic Compounds in the Range δ 0–100

and are influenced by the deshielding or shielding effects of atoms and groups in a manner similar to that of protons. Perhaps surprisingly, however, the chemical shifts of ^{13}C atoms do not necessarily match the relative positions of the corresponding protons in the ^{1}H spectrum.

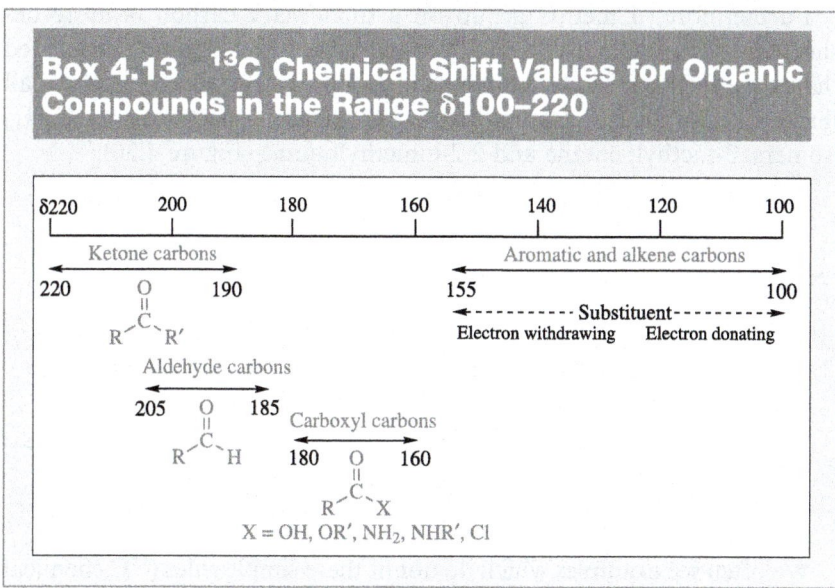

Box 4.13 ^{13}C **Chemical Shift Values for Organic Compounds in the Range δ100–220**

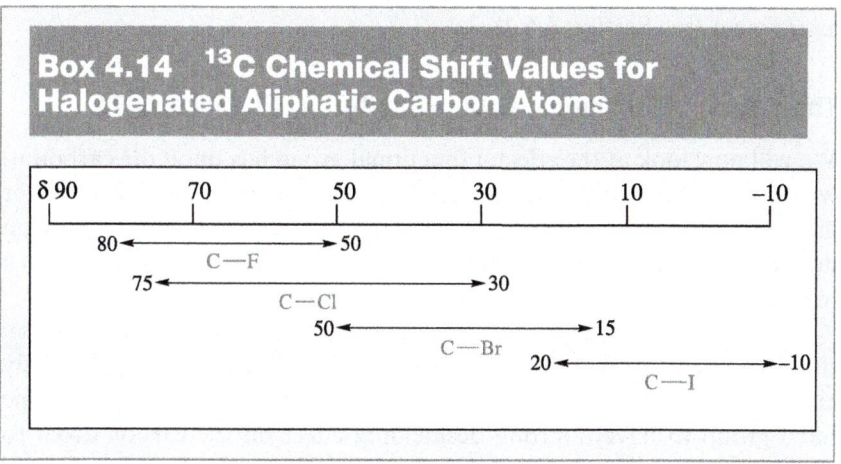

Box 4.14 ^{13}C **Chemical Shift Values for Halogenated Aliphatic Carbon Atoms**

Alkanes

The signals of carbon atoms in an alkyl chain appear at the high-field end of the spectrum, typically about δ 10–30. In general, methyl groups (CH_3) are most shielded (lowest δ value), then methylene (CH_2) and methine (CH), with quaternary alkyl carbon atoms (no attached Hs) being most deshielded (highest δ value). Whatever the atom or group attached to an alkyl carbon, we find that *the greater substitution of that carbon, the more deshielded its signal.* Thus, we nearly always find methyl carbons at lower δ values than the corresponding methylene, methine or quaternary carbon atoms.

Furthermore, a methyl group on a quaternary carbon is more deshielded than a methyl attached to a methine, which is more deshielded than a methyl attached to a methylene group. We can see examples of all three of these methyl groups in the ^{13}C chemical shifts of two C_6H_{14} isomers, 3-methylpentane and 2,2-dimethylbutane (Figure 4.30).

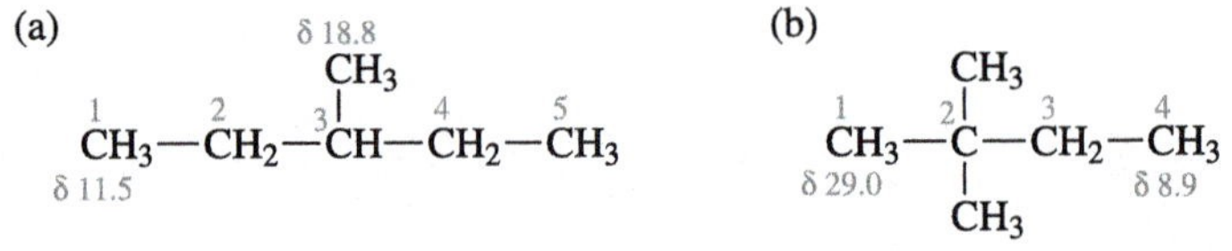

We normally only quote ^{13}C chemical shifts to one decimal place, except those with a five in the second decimal place, *e.g.* δ 43.65.

Figure 4.30 ^{13}C chemical shifts of methyl substituents of (a) 3-methylpentane and (b) 2,2-dimethylbutane

We often see examples which do not fit these simple rules (^{13}C chemical shifts are too variable to make assignments certain), and it is always advisable to use other complementary NMR techniques to aid spectral assignment (see Section 4.6.3).

The Effects of Functional Groups on the ^{13}C Chemical Shift

We will now look at the effect a functional group has upon the carbon to which it is attached. When we have an electron-withdrawing atom or group on a particular carbon atom, we find a reduction in the electron density, resulting in a deshielding of the carbon and a shift downfield (to a higher δ value).

To illustrate this theory, consider the effect of the nitro group (NO_2) on the carbon chemical shifts in nitroethane, $CH_3CH_2NO_2$. The positively charged nitrogen is strongly electron withdrawing and we can expect the nitro group to have a strong deshielding effect on the carbon atom to which it is attached. The fully decoupled ^{13}C spectrum of nitroethane (Figure 4.31) has two signals, one at δ 70.75 and the other at δ 12.3. The methylene carbon (CH_2) is strongly deshielded by the electron-withdrawing nitro group and appears at δ 70.75. This deshielding effect is not transmitted significantly through σ-bonds, and we find the methyl group within the usual alkyl range at δ 12.3. You might like to consider what gives rise to the triplet (at δ 77.0).

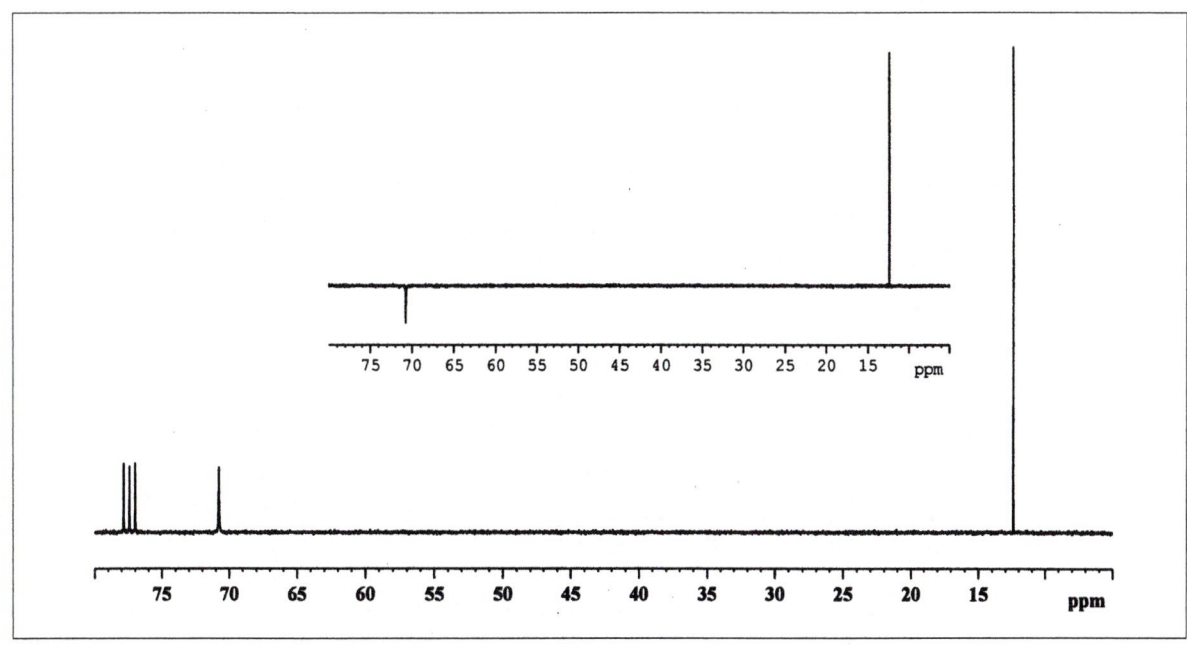

Figure 4.31   75 MHz proton-decoupled ^{13}C NMR spectrum of nitroethane, $CH_3CH_2NO_2$, and DEPT 135 spectrum (*inset*) in $CDCl_3$

Perhaps surprisingly, an electron-donating atom or group on a carbon atom also results in a decrease in electron density on that particular carbon atom and a shift of the signal to lower field. We can partly explain this effect in terms of the electronegativity of most electron-releasing groups, *e.g.* OR or NHR, and the carbon atom directly bonded to the heteroatom "feels" only this electron-withdrawing nature. For example, the methylene bonded to the N in ethylamine, $CH_3CH_2NH_2$, gives rise to a signal at δ 36.7, while the corresponding carbon in ethanol, CH_3CH_2OH, appears at δ 57.8. Both nitrogen and oxygen exert an electron-withdrawing effect on the attached sp^3 hybridized carbon, causing deshielding of the signal, but oxygen, being more electronegative than nitrogen, has a stronger deshielding effect. The methyl groups of ethylamine and ethanol are less affected, and appear at δ 18.8 and δ 18.1, respectively.

There are not many atoms or groups that lead to shielding of the attached carbon, but the nitrile (–C≡N), ethynyl (acetylenic, –C≡C–) and iodo (and to a lesser extent, bromo) groups do have this effect, so in propanenitrile (propionitrile), but-1-yne and iodoethane we find the methylene carbon atom upfield relative to most other methylene groups (Table 4.1). In these compounds the shielding effect is possibly due to the large cloud of electrons on the substituent being close enough to lead to an increase in electron density and so in the shielding of the attached carbon.

Table 4.1 ^{13}C chemical shifts of methylene groups in substituted ethanes in which the methylene carbon is shielded

$CH_3\underline{C}H_2X$	Propanenitrile $X = C\equiv N$	But-1-yne $X = C\equiv CH$	Bromoethane $X = Br$	Iodoethane $X = I$
δ_C	10.9	12.3	27.9	−1.05

Alkenes and Alkynes

When we look at alkenes (sp^2 hybridized) and alkynes (sp hybridized), we find that they are deshielded compared to alkane carbon atoms (sp^3 hybridized). Alkene carbons generally appear between δ 100–140, while alkyne carbons are found at δ 75–105; however, in both these systems the exact chemical shift depends upon the substituents present on the multiple bonds. Similar to the effects upon alkanes, both electron-withdrawing and -donating substituents (Z) directly attached to an alkenic or alkynic carbon (C_α) lead to deshielding of that carbon signal, compared to that of ethene or ethyne (Figure 4.32). However, the chemical shift of the adjacent alkenic or alkynic carbon atom (C_β) is affected differently by electron-withdrawing and electron-donating groups.

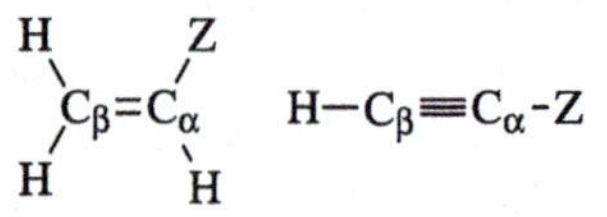

Figure 4.32 Substituted alkenes and alkynes

Electron-withdrawing groups, such as aldehydes and carboxylic acids, cause deshielding to both C_α and C_β, so that C_β is usually more deshielded, while electron-donating groups cause deshielding of C_α but shielding of C_β.

If we look at the ^{13}C spectrum (proton decoupled) of acrylic acid (propenoic acid) (Figure 4.33), we find three unique carbon signals at δ 128.1, 133.15 and 172.0. We can easily identify the carboxyl carbon signal at δ 172.0 as it has a double bond to the highly electronegative oxygen; this leaves the two alkenic, C2 (or α) and C3 (or β), carbon signals, which appear at δ 128.1 and δ 133.15. The O atom, with its high electronegativity, polarizes the C=O bond so that the carbonyl carbon is electron deficient ($\delta+$). In a conjugated unsaturated system like this, the π-bond of the C=C will be polarized by the presence of the electron-withdrawing carbonyl group so that C2 (C_α) will have a slightly greater electron density than C3 (C_β). In summary, both C_α and C_β are deshielded due to the strong electron-withdrawing effect of the carboxyl group, but C_β is more affected than C_α. Such effects can be relayed along an extended

As used here, $\delta+$ and $\delta-$ do not refer to chemical shifts but to partial positive and negative charges.

conjugated system, with each double bond being polarized so that we obtain alternating $\delta+$ and $\delta-$ carbon atoms.

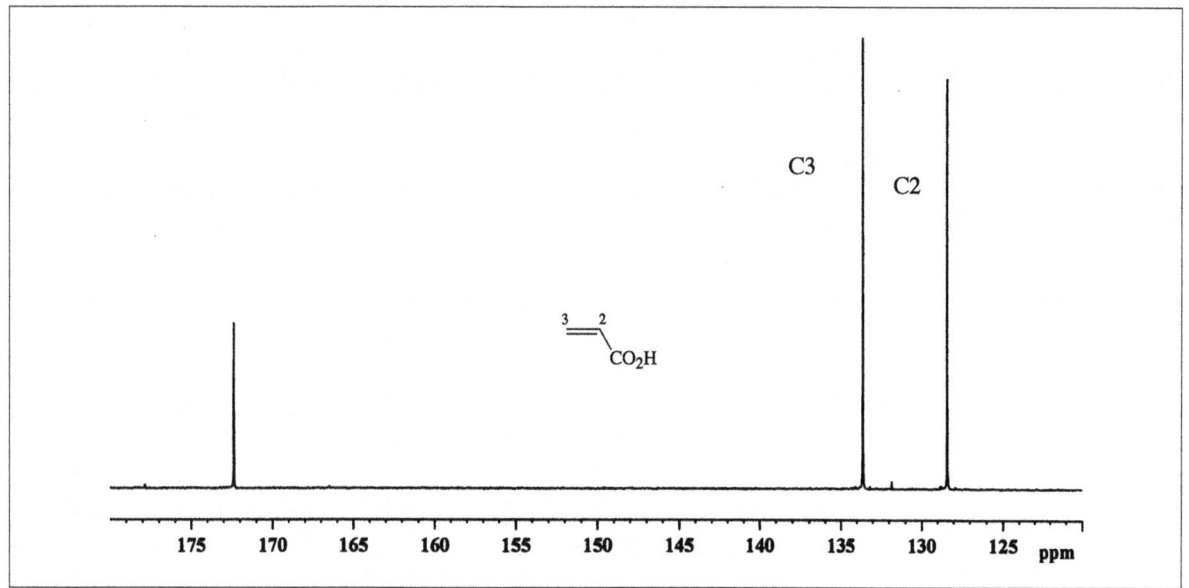

Figure 4.33 Proton-decoupled 75 MHz ^{13}C NMR spectrum of acrylic acid in CDCl$_3$

An exactly opposite effect occurs in alkenes substituted by electron-donating groups, in which C_α is deshielded and C_β shielded. In general, both electron-withdrawing (EWG) and electron-donating groups (EDG) affect the electron density on C_α and C_β of an alkene, with the effect usually being greater on C_β (Figure 4.34).

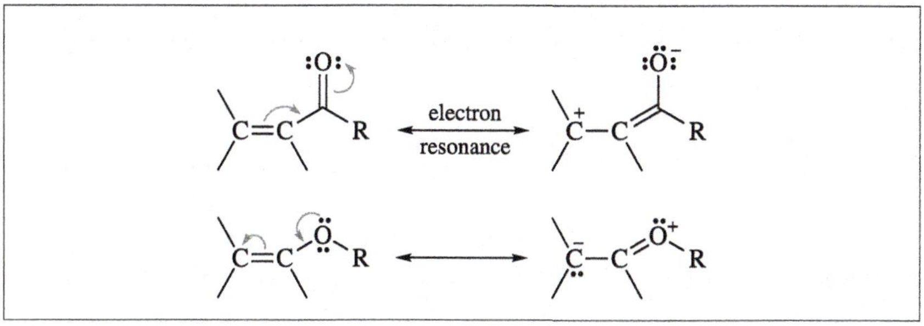

Figure 4.34 The effect of electron withdrawing (*e.g.* COR) and donating (*e.g.* OR) groups on the electron density of alkene carbon atoms

Aromatic Systems

Aromatic carbon atoms usually give rise to signals in the region δ 120–140, although the exact chemical shift is affected by the substituents on the ring (benzene itself has a ^{13}C chemical shift of δ 128.3).

Atoms or groups that can withdraw electrons from (EWG) or donate electrons to the π-system (EDG) lead to different patterns of deshielding or shielding of the ring carbon atoms. The size of the shielding or deshielding effect depends upon the amount of electron donation or withdrawal by the attached group. As we have already seen with alkanes and alkenes, most substituents, whether electron donating or electron withdrawing, lead to some deshielding of the carbon to which they are attached (*ipso*). The more electronegative the attached atom, the more deshielded the substituted carbon atom. As before, the exceptions are $-C\equiv N$, $-C\equiv CH$, $-Br$ and $-I$, which all have a shielding effect.

We can predict whether the 2- (*ortho*), 3- (*meta*) and 4- (*para*) carbon atoms are shielded or deshielded by a particular substituent using the rule of thumb given in Box 4.15.

Box 4.15   Effect of EWG and EDG on the ^{13}C Chemical Shifts of Aromatic Carbon Atoms

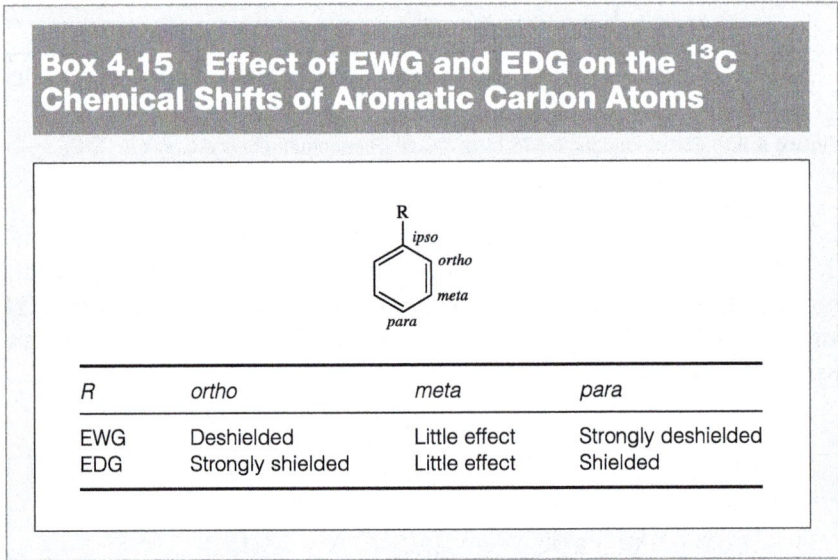

R	ortho	meta	para
EWG	Deshielded	Little effect	Strongly deshielded
EDG	Strongly shielded	Little effect	Shielded

We can understand these substituent effects better by considering the resonance forms possible for an aromatic system with an electron-withdrawing group (*e.g.* C=O) or an electron-donating group (*e.g.* OH) (Figure 4.35). We can see immediately from the resonance structures that the most significant effects are seen at the 2- (*ortho*) and 4- (*para*) positions, whether the substituent is electron donating or withdrawing and, in practice, the 3- (*meta*) carbon is rarely affected by more than a few ppm.

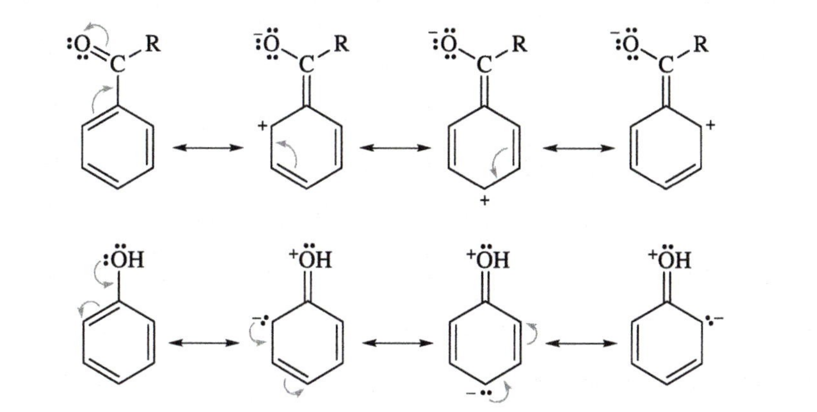

Figure 4.35 Resonance forms for aromatic molecules with C=O or OH groups

In a manner analogous to that used previously to predict ^1H chemical shifts in aromatic systems (Box 4.6), we can use substituent constants to estimate the ^{13}C chemical shifts of a range of compounds, *e.g.* substituted aromatic compounds (Box 4.16).

Box 4.16 Aromatic ^{13}C Chemical Shifts

$$\delta_C = 128.3 + \sum \Delta_X$$

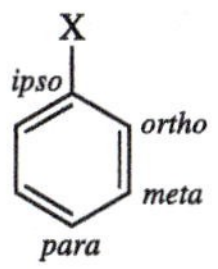

Substituent, X	Carbon chemical shift (δ)			
	Δ_{Xipso}	Δ_{Xortho}	Δ_{Xmeta}	Δ_{Xpara}
Me	9.5	0.8	0	−2.9
Ph	12.5	−1.5	0.1	−1.5
CH$_2$OH	12.6	−1.3	0.1	−0.9
COR	8.8	1.1	0.3	5.0
CO$_2$R	1.6	1.7	0.2	5.1
CH=CH$_2$	9.4	−2.1	0.2	−0.5
C≡CH	−6.1	3.8	0	0.4
C≡N	−15.9	3.8	0.9	4.5
OR	29.0	−13.6	1.4	−7.4
SH	2.5	0.7	1.0	−2.8
NH$_2$	18.2	−13.2	1.0	−9.9

(continued)

NHCOMe	9.9	−7.9	0.5	−4.1
NO$_2$	20.0	−4.8	1.1	6.4
F	34.8	−12.8	1.8	−4.1
Cl	6.0	0.3	1.4	−1.9
Br	−5.8	3.2	1.7	−1.5
I	−33.9	9.0	1.8	−1.0

Data taken from SDBS website (http://www.aist.go.jp/RIODB/SDBS/menu-e.html). All spectra used for these values were recorded in CDCl$_3$. The chemical shift of benzene in CDCl$_3$ is 128.3.

Carbonyl Groups

Carbonyl carbons, such as those in esters, carboxylic acids, amides, ketones and aldehydes, in which the sp^2 carbon is bonded by a π-bond to the highly electronegative O, have a characteristic chemical shift of δ 160–220 (Box 4.13). Within this range, the sp^2 carbon atoms of aldehydes and ketones are most deshielded and usually resonate at δ 185–220, while those of carboxylic acids and their derivatives resonate about δ 160–180.

Worked Problem 4.1

Q Assign the signals for the proton-decoupled ^{13}C spectrum of paracetamol (Figure 4.36).

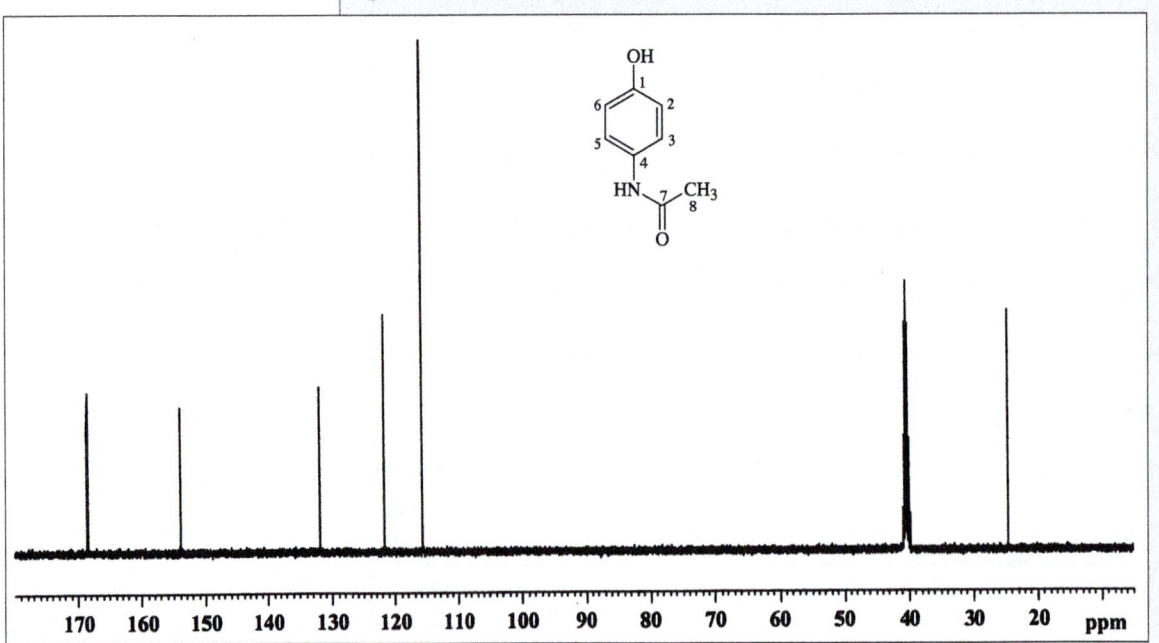

Figure 4.36   125 MHz proton-decoupled ^{13}C spectrum of paracetamol in DMSO-d_6

A We see examples of many of the points raised about chemical shifts when we consider this spectrum. Starting with the most extreme signals, we can assign the most deshielded signal, at δ 168.4, to the amide (C7) and the most shielded, at δ 24.6, to the methyl (C8). There is a plane of symmetry in this aromatic ring, such that C2 and C6 are chemically and magnetically equivalent and so give rise to only one signal for the pair, as are C3 and C5, which also give rise to a single signal. A first guess would suggest that these are the higher intensity signals at δ 121.7 and 115.9. Although both OH and NHR are examples of EDGs, the nitrogen lone pair is involved in resonance with the amide carbonyl and is not readily available for resonance with the aromatic ring, so is less electron donating than the OH (see Figure 4.7).

Remembering the effect electron resonance of an EDG group has upon the ring, we can predict that the *ortho* (C2 and C6) and *para* (C4) carbons to the OH will have highest electron density and will be most shielded, and we see the signal for C2/C6 at δ 115.9. The signal at δ 121.7 is therefore due to C3/C5, which are in the *ortho* position to the mildly electron-donating NHCO group and are slightly shielded (the electronegative oxygen has little effect on the chemical shift of the C3/C5 carbons which are *meta* to it).

C4 is directly attached to nitrogen and, for this reason, it is deshielded (slightly compensated by the *para* electron-donating oxygen) and appears at δ131.9, while the signal at δ 154.0 for C1 is more strongly deshielded than C4 as it is attached to the strongly electronegative oxygen. Notice that both these quaternary carbon atoms have a lower intensity than the CH atoms owing to their poor relaxation rates.

Some characteristic chemical shift ranges were given in Boxes 4.12–4.14. However, the exact chemical shift of a particular carbon atom can be affected by a range of factors, such as temperature and conformational restriction, so that chemical shift is not usually a key factor in the assignment of ^{13}C NMR spectra.

4.6.2 Coupling to ^{13}C Nuclei

^{13}C exhibits coupling to other NMR-active nuclei, such as ^{1}H and ^{2}H (D), ^{19}F and ^{31}P. The coupling arises in the same way as ^{1}H–^{1}H coupling, which we met earlier in Section 4.4.2, and you can use the same arguments to predict and explain the observed coupling to ^{13}C. We can use the coupling from ^{13}C to other atoms to help us in the interpretation of spectra.

Carbon-13 ($^{13}C-^{13}C$) Coupling

$^{13}C-^{13}C$ coupling is not usually observed in the spectra of organic molecules, as there is only about 0.01% chance ($\equiv 1.1\%$ of 1.1%) of finding two ^{13}C atoms adjacent to one other in the same molecule, and the resulting signal is too weak to be observed. Of course, if a sample is enriched in ^{13}C, as would be found in a ^{13}C-labelled (enriched) compound, then $^{13}C-^{13}C$ coupling is seen, and is usually of the order of 40–180 Hz, with larger J values for unsaturated carbons atoms owing to the shorter bonds in these systems.

Proton ($^{1}H-^{13}C$) Coupling

Decoupled Spectra

The major coupling to carbon that can be detected in any organic molecule is the $^{1}H-^{13}C$ coupling. The presence of this coupling can complicate the ^{13}C spectrum dramatically, so it is usually removed by irradiating over the entire proton frequency range while obtaining the carbon spectrum. This gives rise to the standard proton-decoupled ^{13}C spectrum (often called broad-band decoupled spectrum), in which each non-equivalent carbon atom gives rise to a single peak. A further advantage of this irradiation of all the protons in the molecule is that the carbon spectrum benefits from the nuclear Overhauser effect (NOE) and the signal-to-noise ratio in the carbon spectrum is improved (by as much as 300%). Organic chemists most often request this type of ^{13}C spectrum as it allows rapid assignment of the majority of the carbon signals, and we have already seen an example when we assigned the ^{13}C spectrum of paracetamol (Figure 4.36).

Undecoupled and Off-resonance Spectra

Sometimes the coupled ^{13}C spectrum is useful in determining whether a carbon atom is a quaternary (no attached Hs), methine (CH), methylene (CH$_2$) or methyl (CH$_3$). Like $^{1}H-^{1}H$ coupling, $^{1}H-^{13}C$ coupling follows the $(n+1)$ rule, so that the carbon atom of a methyl group appears as a quartet, that of a methylene appears as a triplet, that of a methine appears as a doublet and quaternary carbons are seen as singlets. Such spectra are usually referred to as undecoupled spectra and $^{1}J_{CH}$ is usually of the order of 120–250 Hz, and is dependent upon the amount of unsaturation. $^{2}J_{CH}$ and $^{3}J_{CH}$ coupling constants are usually less than 20 Hz, except for ^{2}J coupling from a carbonyl carbon to a *vicinal* hydrogen, $-\underline{C}(=O)-C-\underline{H}$, which is about 30 Hz, and from C2 of a terminal alkyne to the terminal hydrogen atom, $-\underline{C}\equiv C-\underline{H}$, which is about 50 Hz.

Undecoupled spectra are rarely obtained since $^{1}J_{CH}$ coupling constants are so large that very complex spectra can arise (made even more complex by $^{2}J_{CH}$ and even $^{3}J_{CH}$). In order to simplify the spectra, partial decoupling of the ^1H–^{13}C signals produces an off-resonance spectrum, in which only the $^1J_{CH}$ coupling is seen. These spectra require a reduced acquisition time compared to undecoupled spectra, but have limited usefulness as the outer peaks in multiplets are sometimes lost in the baseline noise, making it difficult to distinguish between quartets and doublets, triplets and singlets (Figure 4.37).

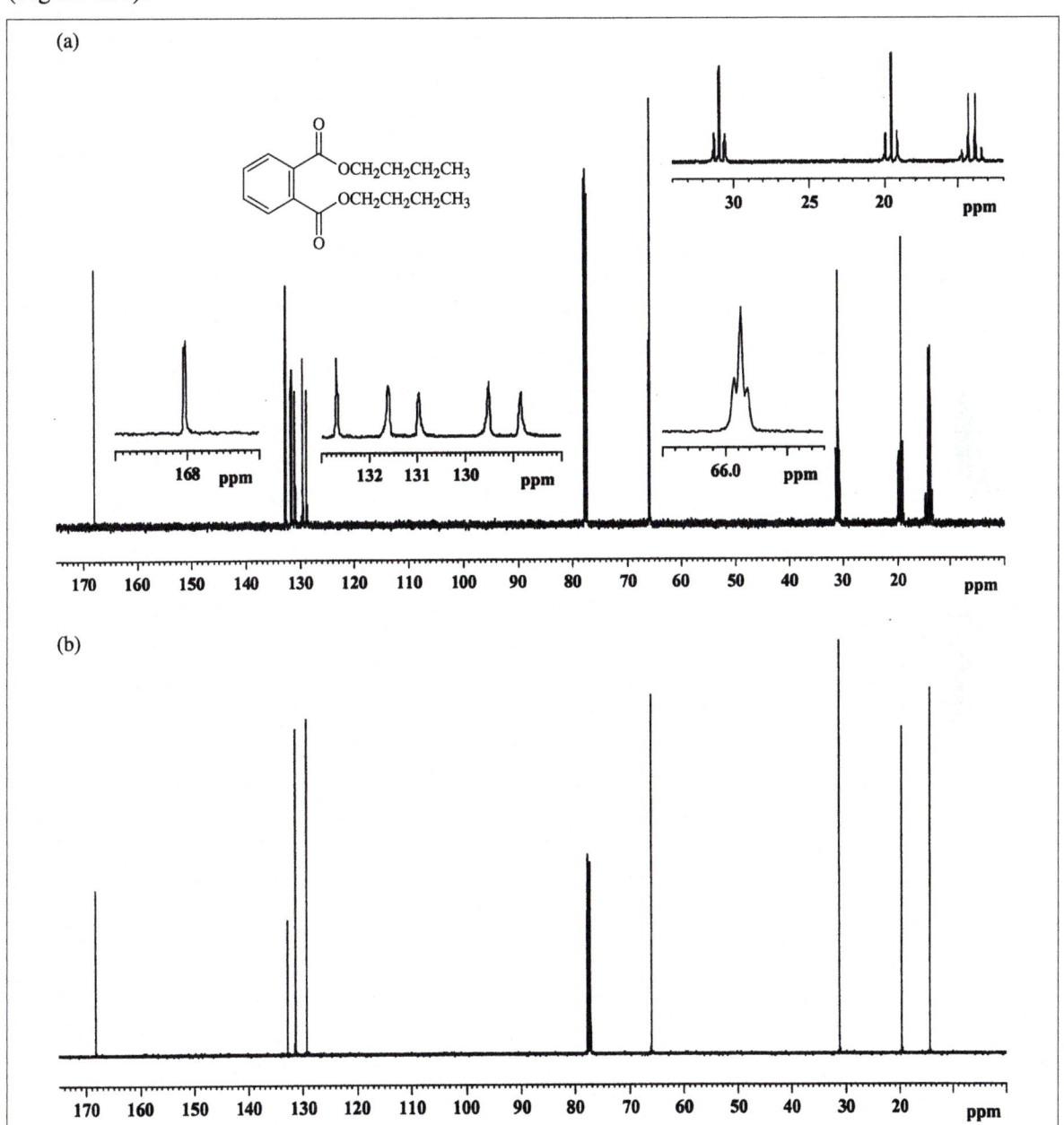

Figure 4.37   (a) Off-resonance decoupled 125 MHz ^{13}C spectrum of dibutyl phthalate in $CDCl_3$; (b) proton decoupled 125 MHz ^{13}C spectrum of dibutyl phthalate in $CDCl_3$

Deuterium (^{2}H-^{13}C) Coupling

Deuterium (^{2}H, more commonly denoted D) has a spin (I) of 1 and D–^{13}C coupling is a common feature of ^{13}C NMR spectra (see Box 4.17 for prediction of peak multiplicities), since they are normally obtained using CDCl$_3$ or DMSO-d_6 as the solvent (see the peaks at δ 77.0 in Figure 4.37 and δ 40.5 in Figure 4.36, respectively). Deuterium has a natural abundance of 0.015%, so D–^{13}C coupling is only really seen in D-labelled compounds and its decoupling is rarely necessary. The size of the D–^{13}C coupling constant is about ⅙ that of the corresponding ^{1}H–^{13}C coupling.

Box 4.17   Prediction of Peak Multiplicities

We can predict the multiplicity of a particular coupled signal using the relationship:

$$\text{Multiplicity} = (2 \times I \times \text{number of coupling atoms}) + 1$$

so that, in the case of CDCl$_3$, the multiplicity of the carbon will be: $(2 \times 1 \times 1) + 1 = 3$, *i.e.* a triplet. In DMSO-d_6, in which there are two identical CD$_3$ groups, this gives rise to a septet: $(2 \times 1 \times 3) + 1 = 7$.

Fluorine (^{19}F–^{13}C) and Phosphorus (^{31}P–^{13}C) Coupling

Both ^{19}F and ^{31}P have 100% natural abundance and a spin (I) of ½, meaning that, if present in a molecule, they will also give rise to coupling with ^{13}C and ^{1}H within the usual bond ranges. $^1J_{CF}$ is usually between 150–180 Hz for aliphatic compounds, while $^2J_{CF}$ and $^3J_{CF}$ are generally between 5 and 25 Hz. Slightly larger ^{19}F–^{13}C couplings are found in aromatic systems ($^1J_{CF}$ is about 250 Hz), which also exhibit longer range ^{19}F–^{13}C coupling, with $^4J_{CF}$ of the order of 3 Hz.

Organophosphorus compounds rarely have more than one P atom in the molecule, giving rise to a doublet for any C atom within three bonds (assuming the proton–carbon coupling has been removed) and so these compounds show clearly the connectivity of carbon atoms close to the P atom. The size of the coupling constant is dependent on the number of bonds and the phosphorus oxidation state, such that ^1J is the largest at about 45–150 Hz, while ^2J and ^3J are of the order of 10–15 Hz.

4.6.3 ^{13}C Assignment Techniques

As mentioned earlier, the irradiation of directly bonded hydrogen atoms increases the intensity of ^{13}C signals through the NOE, which means that CH, CH$_2$ and CH$_3$ groups usually have a relatively high intensity in a ^{13}C NMR spectrum, while we can often recognize quaternary carbon atoms because of their low intensity. This difference in intensity is due to the different relaxation rates, and is the reason we do not integrate ^{13}C NMR spectra.

There are many assignment techniques which can be run to aid interpretation and assignment of difficult ^{13}C NMR spectra, with the most commonly used being:

- Distortionless Enhancement by Polarization Transfer (DEPT).
- Heteronuclear Multiple Quantum Correlation (HMQC).
- Heteronuclear Multiple Bond Correlation (HMBC).

Used together, these spectra have revolutionized structure elucidation by helping organic chemists to deduce the connectivities of hydrogens to the corresponding carbon atoms.

DEPT Spectra: Identifying Quaternary, Methine, Methylene and Methyl Carbons

DEPT uses a pulse sequence that includes a delay between the excitation pulse and the detection of the emission signal from the carbon atoms which is related to the C–H coupling constant, in order to distinguish between the different types of carbon atom (we do not need to understand how this is achieved).

As DEPT relies on the transfer of polarization from a directly bonded H atom to the carbon – resulting in the increased sensitivity of the carbon atoms – only ^{13}C atoms that are attached to H atoms are detectable by this method, and so no quaternary carbon atoms are seen on DEPT spectra. Depending upon something called the pulse angle (which is expressed as a number after the acronym, but we do not need to know its significance), there are three different DEPT experiments that can be carried out on a particular sample.

DEPT 135 gives positive signals (vertically up from the baseline) for methyl (CH$_3$) and methine (CH) carbon atoms; methylene (CH$_2$) signals are negative and appear below the baseline.

Although quaternary carbons do not appear on DEPT spectra, we can identify them by comparing DEPT and proton decoupled spectra.

DEPT 90 only detects the methine carbons which are seen as positive signals.

DEPT 45 is rarely used, as it is less differentiating, with all methine, methylene and methyl signals being positive.

DEPT spectra are particularly useful in assigning aliphatic parts of molecules, where there may be methine, methylene and methyl groups all fairly close in chemical shift; they are also of great help when two carbon signals coincide, providing they are seen in opposite domains, *e.g.* the methyl and methylene signals in a DEPT 135 spectrum (Figure 4.38).

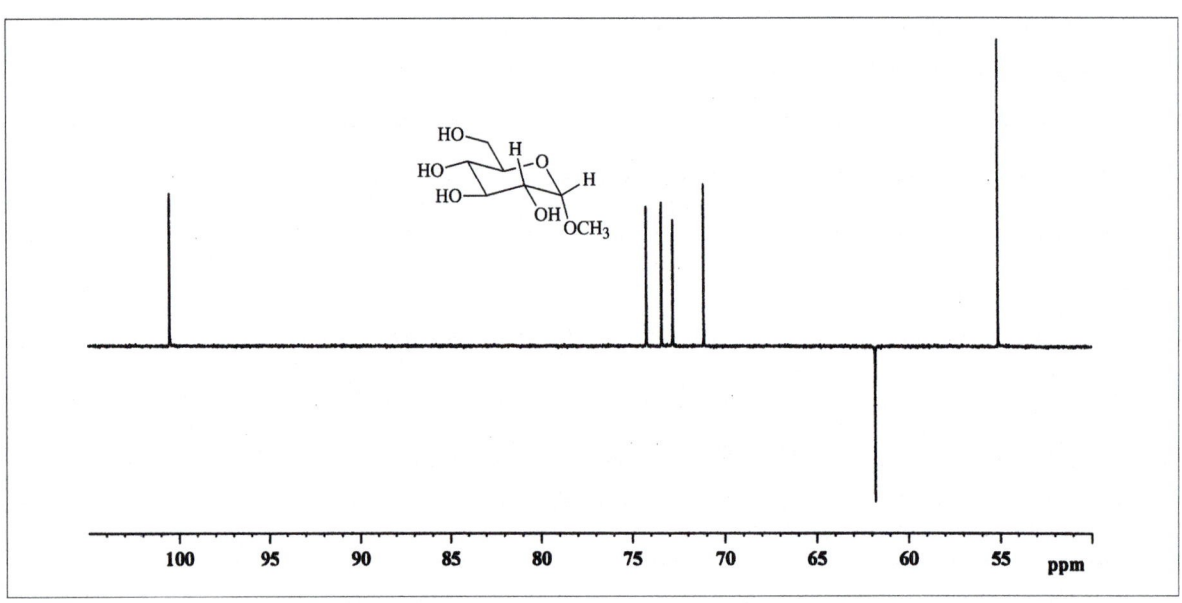

Figure 4.38   75 MHz ^{13}C DEPT 135 spectrum of methyl α-glucopyranoside in DMSO-d_6

Two-dimensional COSY

There are several variations of correlation spectroscopy, all giving rise to different, complementary data. We have already met HH (homonuclear) COSY earlier in this chapter (Section 4.5). An obvious extension to the COSY experiment is to use it for heteronuclear correlation, *e.g.* correlation of all the ^{1}H and ^{13}C signals in a molecule so that we can determine which ^{1}H is attached to which ^{13}C. The CH COSY technique does this, and correlates the ^{13}C chemical shifts on one axis (usually the y-axis) with the ^{1}H shifts on the other (usually the x-axis) through the one-bond (^1J) CH coupling. In the CH COSY spectrum, each ^{1}H–^{13}C pair gives a single cross-peak in the spectrum. The pulse sequence employed (which again we do not need to understand) leads to polarization transfer from the ^{1}H to the ^{13}C, leading to an increase in the ^{13}C sensitivity, but, since this technique involves detection of the carbon nuclei, it requires large amounts of sample and is rather time consuming.

Similar spectra can be obtained more rapidly and with less sample if the data are acquired through the proton signals, which are much more intense. Basically, the ^{1}H NMR data are acquired and the ^{1}H–^{13}C coupling constant used as the delay in a pulse sequence, which enables us to obtain the carbon spectrum. This method of obtaining the data is called "inverse-mode", since the carbon atoms are detected through their attached hydrogen atoms rather than by direct detection, with obvious benefits in the sensitivity and the time taken to obtain a spectrum. HMQC and HMBC are both examples of "inverse-mode" spectra and this method is so much quicker than CH COSY that an entire HMQC spectrum can be obtained in much less time than it takes to obtain the proton-decoupled ^{13}C spectrum.

HMQC Spectra

Like COSY spectra, the HMQC experiment gives rise to a two-dimensional spectrum that shows the correlation between a carbon atom and the attached proton(s), $i.e.$ it shows $^1J_{CH}$ coupling. The carbon signals are detected through the low-intensity satellite signals (due to ^{1}H–^{13}C coupling) in the proton spectrum, which enables the chemical shift of the coupled carbon to be extrapolated. Since the carbon atoms are detected through the attached protons, quaternary carbons are again not detected in this experiment. The proton spectrum is usually plotted along the x-axis for reference and the carbon chemical shifts are usually plotted along the y-axis (the ^{13}C spectrum is often not plotted). The carbon–proton correlations show as contours (cross-peaks) at the intersection of each of the signals (Figure 4.39). One major advantage of this technique is that, since the ^{13}C spectrum is spread over 200 ppm, even overlapping ^{1}H signals are separated when correlated with the ^{13}C dimension, as shown for the multiplets in Figure 4.39.

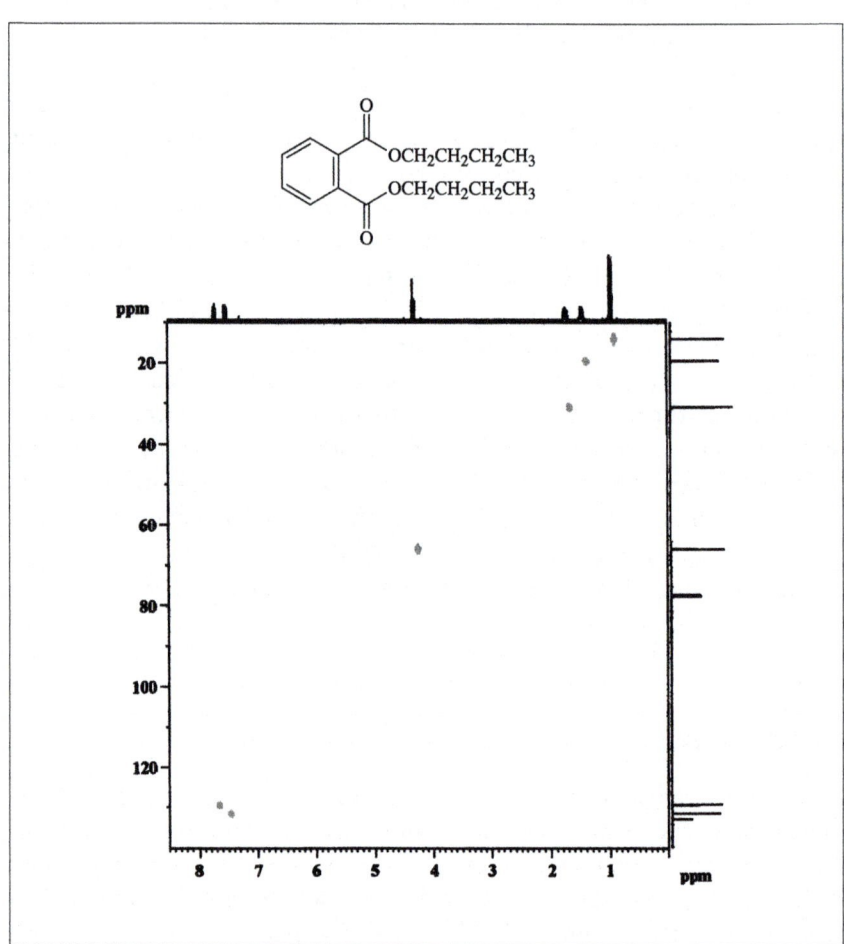

Figure 4.39 HMQC [$^1J(^1H–^{13}C)$] spectrum of dibutyl phthalate in CDCl$_3$

The application of this technique can be seen in Figure 4.40, in which the HMQC spectrum of paracetamol is shown. In this spectrum, all the carbon atoms with directly attached protons are immediately assignable: the methyl carbon can be assigned to the signal at δ 24.6, C2 and C6 to that at δ 115.9, and C3 and C5 to that at δ 121.7. Using this method, we have confirmed our earlier assignments and can be fairly certain of their accuracy.

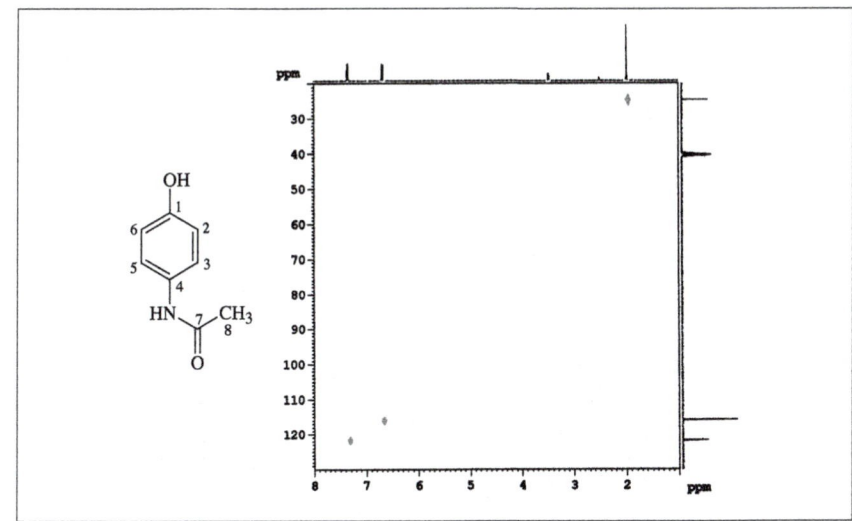

Figure 4.40 HMQC [$^1J(^1\text{H}-^{13}\text{C})$] NMR spectrum of paracetamol in DMSO-d_6

HMBC Spectra

Sometimes, however, the exact arrangement of the carbon atoms is not immediately clear from the standard spectra and, of course, we cannot identify the quaternary carbon atoms from an HMQC spectrum. In these cases, further information is required about atom connectivities, and as ^{13}C is so low in natural abundance, a $^{13}\text{C}-^{13}\text{C}$ correlation experiment is not realistically possible with unlabelled compounds. It is possible, however, to detect $^1\text{H}-^{13}\text{C}$ couplings over two bonds (giving rise to $^2J_{\text{CH}}$ coupling) and even three bonds ($^3J_{\text{CH}}$ coupling); in certain molecules, $^4J_{\text{CH}}$ coupling can even be seen (usually in unsaturated systems). The experiment which detects this longer range $^1\text{H}-^{13}\text{C}$ coupling, and once again converts it into a correlation map, is called heteronuclear multiple bond correlation. HMBC spectra are also two-dimensional "inverse mode" spectra and they show the connectivities of carbon atoms through their $^2J_{\text{CH}}$ and $^3J_{\text{CH}}$ coupling. Once again, the correlations are represented by contours at the intersection of the appropriate H and C signals on the 2D spectrum (Figure 4.41).

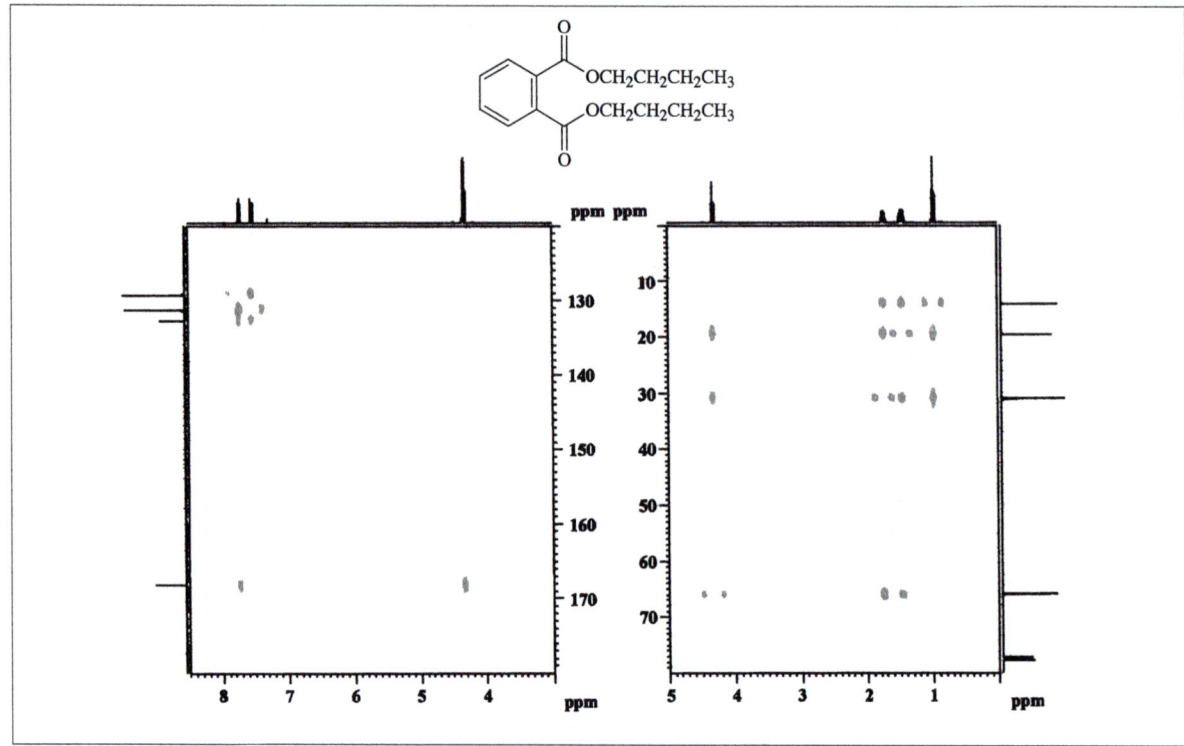

Figure 4.41   HMBC [$^2J(^1H-^{13}C)$ and $^3J(^1H-^{13}C)$)] spectrum of dibutyl phthalate in CDCl$_3$

Using HMBC, the longer range connectivities are detected from a particular H to the carbon atom(s) either two bonds away (H–C–C: ^2J coupling) or three bonds away (H–C–C–C: ^3J coupling), so that, indirectly, the carbon–carbon connectivities can be deduced and a map can be generated of the carbon and hydrogen connectivities.

When HMBC spectra are used in conjunction with HH COSY spectra, the $^2J_{CH}$ couplings can be identified in the HMBC spectrum because the vicinal hydrogen atoms can be identified from the COSY spectrum and related to their attached carbon atoms.

If we return to paracetamol to examine the HMBC spectrum, we can see how useful this experiment is. Remember, the HMBC spectrum shows ^2J and ^3J CH coupling. This means that a cross-peak vertically below a proton signal indicates that the proton is two or three bonds away from the carbon atom to which that signal corresponds.

When the structure of the compound under investigation is known, or at least suspected, the interpretation process can be aided by devising a table of expected ^2J and ^3J CH coupling. We can then compare the expected cross-peaks to the observed peaks to confirm, or otherwise, the structure. Such a table for paracetamol would appear as in Table 4.2.

Table 4.2 Expected ^2J and ^3J CH correlations for paracetamol[a]

H (δ) C (δ)	NH (9.66)	OH (9.14)	3/5 (7.34)	2/6 (6.69)	CH$_3$ (1.99)
8 (24.6)	*				
2/6 (115.9)		*		(*)	
3/5 (121.7)	*		(*)	•	
4 (131.9)	•		•	*	
1 (154.0)		•	*	•	
7 (167.4)	•				•

[a] • represents an expected $^2J(^1H-^{13}C)$ correlation and * a $^3J(^1H-^{13}C)$ correlation; (*) represents a possible $^3J(^1H-^{13}C)$ correlation from H2 to C6 or from H3 to C5, *etc.*

Worked Problem 4.2

Q Using both the full HMBC spectrum (Figure 4.42) and the expanded section of this spectrum, compare the cross-peaks with those expected (Table 4.2) and so confirm that almost all the expected cross-peaks can be seen.

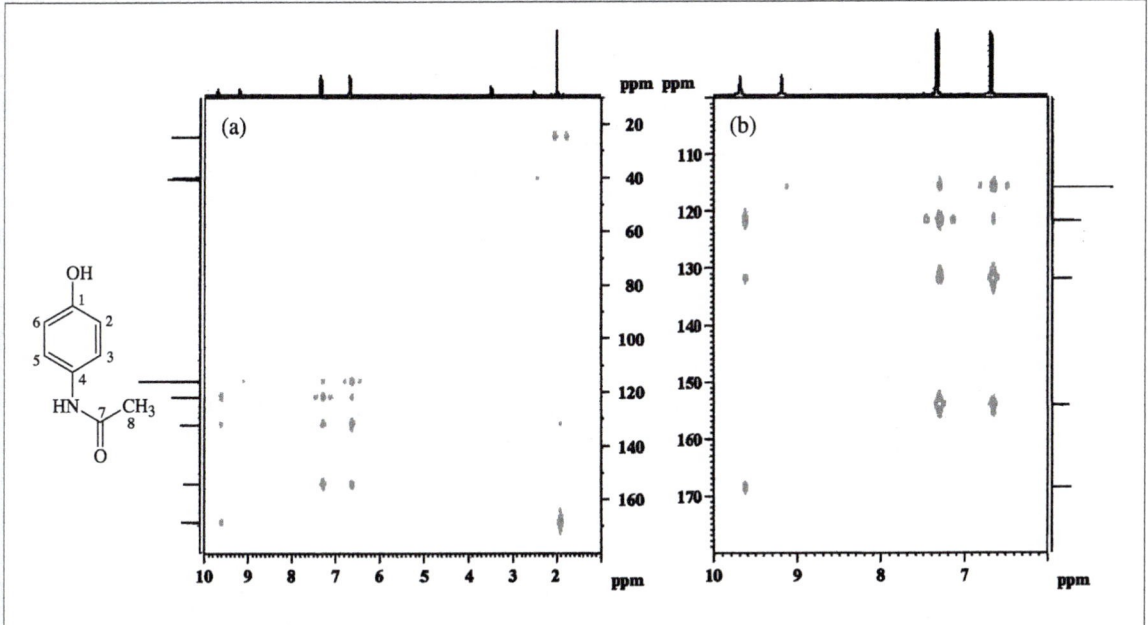

Figure 4.42 (a) HMBC [$^2J(^1H-^{13}C)$ and $^3J(^1H-^{13}C)$] spectrum of paracetamol in DMSO-d_6 and (b) expansion of the aromatic region

A Examining the spectrum, we can see all the expected cross-peaks except some of those to exchangeable protons (OH and NH); we saw earlier that some of the coupling to such exchangeable protons is often absent. To fully evaluate the ^2J and ^3J CH couplings to C2/C6, C3/C5 and C4, the expanded HMBC spectrum is required. In this spectrum the sensitivity has been increased so that even the weaker couplings are observed and all the expected cross-peaks can be seen. We also see the symmetrical ^1J cross-peaks either side of the H2 and H3 signals, but we can ignore these here. The other, perhaps unexpected, cross-peaks are found between H2/H6 and C2/C6, and between H3/H5 and C3/C5. At first, this seems strange, but we have to remember that although H2 is attached to C2 it is not attached to C6: from the point of view of H2, C6 and C2 are not equivalent, such that H2 shows ^1J coupling to C2 but ^3J coupling to C6. The same explanation applies to the cross-peaks seen for H6 (with C2), H3 (with C5) and H5 (with C3).

4.7   Other Nuclei

We have seen earlier in this chapter that certain nuclei, other than carbon and hydrogen, also have nuclear spin and are NMR-active. The fact that there are other NMR-active nuclei leads to the question: can we run spectra specifically to see these other nuclei? The answer is *yes*. In Table 4.2 we can see which nuclei have spin and are NMR-active so that, in theory, the NMR spectrum of any of these nuclei can be recorded. The nuclei of greatest interest to organic chemists, besides ^{1}H and ^{13}C, are ^{14}N (and ^{15}N), ^{17}O, ^{19}F and ^{31}P. Nuclei with 100% abundance, ^{19}F and ^{31}P, present little problem to the NMR spectroscopist and recording the NMR of these nuclei is fairly routine.

Although you would expect deuterium (D) NMR to be every bit as useful as ^{1}H NMR, it is not carried out routinely as deuterium has a natural abundance of 0.015% and very low sensitivity. Furthermore, as the spin is greater than ½, it is a quadrupolar nucleus and the signals are usually broad. Deuterium shows signals in the same chemical shift range as ^{1}H (δ 0–12), and NMR is most often used in experiments using D-labelled compounds to investigate reaction mechanisms.

Although ^{14}N is 99.6% abundant and has spin $I = 1$, it too is a quadrupolar nucleus and this again causes severe line broadening, which makes the signals difficult to observe clearly. ^{15}N and ^{17}O are relatively low in abundance and of even lower sensitivity than ^{13}C, so these NMR experiments are reserved for special applications.

4.8   Liquid Chromatography NMR

LC-NMR combines high-performance liquid chromatography with an NMR detector, with a key feature being the replacement of the aqueous part of the LC mobile phase with D_2O, while using a fully protonated organic component such as methanol (CH_3OH) or acetonitrile (CH_3CN). LC-NMR requires some means of continuously introducing the LC mobile phase into the NMR spectrometer until the LC peak of interest has eluted from the column. The normal NMR probe is, therefore, replaced by a probe containing a flow cell, the volume of which ranges between 50 and 400 μL, and a stopped-flow technique is employed, which involves using a UV detector (or any other form of detector) to trigger the LC pump to stop at the exact moment the LC peak reaches the flow cell. Once the flow has been stopped with the peak of interest in the flow cell, any of a number of different NMR experiments can then be carried out on the sample, *e.g.* HH COSY.

In an LC-NMR probe the sample coil is much closer to the sample and the limit of detection is therefore decreased from the milligram (mg) to the microgram (μg) scale. In conjunction with being able to obtain a large number of scans on the sample stopped in the flow cell, one of the inherent disadvantages of NMR (*i.e.* poor signal-to-noise) is thus partially overcome. Another way of further enhancing the signal-to-noise ratio is to cool the NMR probe to a very low temperature.

As D_2O is used as one of the solvents, we observe a large HOD peak due to exchange with any water remaining in the mobile phase (Figure 4.43). Coupled with this HOD peak is the even larger signal associated with the organic solvent, which must be suppressed by irradiating the solvent signal with low-level continuous radiation before data acquisition is undertaken. Thus, as in decoupling, the energy levels of the solvent signals are saturated and they are reduced to a level where their effect on the signal-to-noise ratio of the NMR spectrum is greatly reduced. One drawback, however, is that any sample signals that happen to occur at the same chemical shift as the solvent peaks are not observed in the spectrum. It is possible, under these circumstances, to change solvent system and thus move the solvent peaks, and the irradiation frequency, to a different chemical shift.

LC is the abbreviated form of the acronym HPLC (high-performance liquid chromatography).

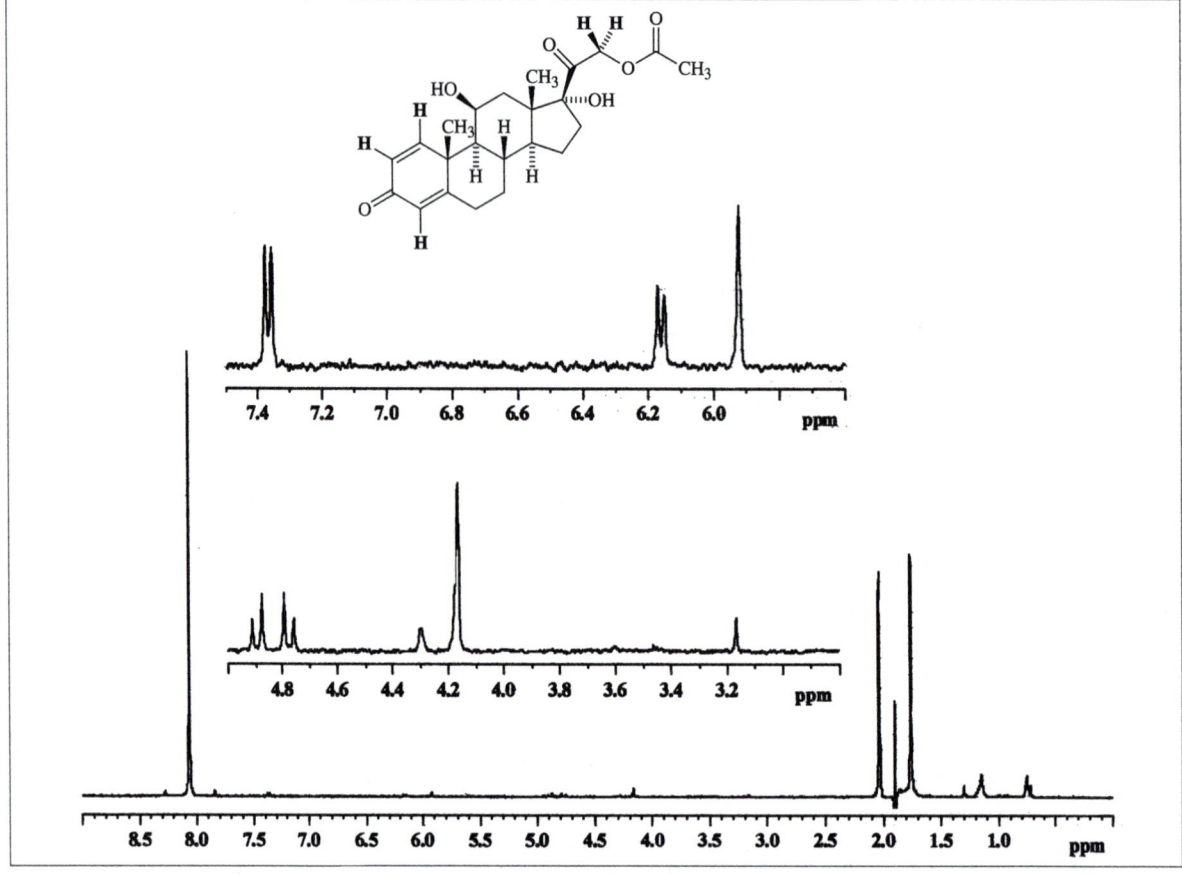

Figure 4.43   500 MHz ^{1}H LC-NMR spectrum of approximately 3 µg of prednisolone 21-acetate in $D_2O/CH_3CN/0.1\%$ HCO_2H, with solvent suppression at δ 1.90 (CH_3CN) and δ 4.16 (residual water). The peak at δ 8.05 is due to HCO_2H. The aliphatic protons are masked by the acetonitrile peak, but the methylene and alkene protons (*highlighted in bold*) can be seen in the range δ 4.5–7.5

Summary of Key Points

1.   Nuclei with $I \neq 0$ can adopt certain allowed orientations in an external magnetic field and NMR transitions (between the energy levels for these orientations) require radiofrequency irradiation. The nuclei can adopt $(2I + 1)$ orientations in the magnetic field and the NMR selection rule states that transitions with $\Delta m_I = \pm 1$ are allowed. The relationship between the magnetic field and the frequency of the transition is given by $v = \gamma B_{\mathrm{eff}}/2\pi$.

2.   Nuclei in molecules are affected by the local electron density and their chemical shift, the ratio of the frequency at which an NMR transition occurs to the strength of the main field as measured by the frequency at which protons resonate, is given by:

$$\delta = \frac{v_{sample} - v_{reference}}{v_{spectrometer}} \times 10^6$$

Tables of chemical shifts can be used to determine the molecular environment of all nuclei, but are especially useful in ^{1}H and ^{13}C NMR spectroscopy.

3.   Peak splitting patterns indicate the number of spin active ($I \neq 0$) nuclei in the molecule which are close to the nucleus. Chemically and magnetically equivalent protons do not couple to one another. Spin–spin coupling is usually restricted in ^{1}H NMR to three bonds between the interacting nuclei and can be predicted by the method of successive splitting, in which we take advantage of the fact that each spin ½ nucleus will split the peak due to its non-equivalent neighbour into a doublet of separation J (Hz). A useful general rule is that if we have n protons coupling to the nucleus of interest, the signal for that nucleus will have $(n+1)$ lines.

4.   The area under each peak in a ^{1}H NMR spectrum is proportional to the number of ^{1}H giving rise to the signal, so that integration of ^{1}H spectra can be used to determine the number of hydrogens in each molecular environment. This is not true for ^{13}C spectra, in which quaternary carbons usually give less intense signals due to inefficient relaxation processes, so that ^{13}C spectra are not normally integrated.

5. ^{13}C spectra give one line for each magnetically distinct carbon atom in a molecule so that assignment of ^{13}C resonances usually requires other techniques, including DEPT and HMQC [$^1J(^1$H–^{13}C)] spectra. Longer range HMBC [$^2J(^1$H–^{13}C) and $^3J(^1$H–^{13}C)] spectra are also useful in structure determination since they provide connectivities through heteroatoms and to quaternary carbon atoms.

Problems

4.1.   The Friedel–Crafts alkylation of benzene with 1-chloropropane in the presence of aluminium trichloride gives, after chromatography, **A** (76%) and **B** (24%). Using the 300 MHz ^{1}H NMR data given below, identify **A** and **B** and, having done so, assign their ^{13}C NMR spectra (*hint*: in both the ^{13}C NMR spectra, the peak at δ 125.7 is smaller than the peaks between δ 126 and δ 129, but easily the smallest peaks in both the spectra are those at δ 142.7 and δ 148.8).

A: δ_H (300 MHz, CDCl$_3$) 1.25 (6H, d, $J = 6.9$ Hz), 2.90 (1H, 7 lines, $J = 6.9$ Hz), 7.18–7.31 (5H, m).

B: δ_H (300 MHz, CDCl$_3$) 0.94 (3H, t, $J = 7.0$ Hz), 1.64 (2H, 6 lines, $J = 7.0$ Hz), 2.57 (2H, t, $J = 7.0$ Hz), 7.02–7.40 (5H, m).

A: δ_C (75 MHz, CDCl$_3$) 24.0, 34.2, 125.7, 126.4, 128.3, 148.8.

B: δ_C (75 MHz, CDCl$_3$) 13.8, 24.6, 38.2, 125.7, 128.3, 128.5, 142.7.

4.2. The ^{1}H (Figure 4.44) and HH COSY (Figure 4.45) spectra of a disubstituted β-hydroxy ketone **C** are shown below. Use these spectra, and your knowledge of chemical shifts and coupling constants, to fully assign the ^{1}H spectrum.

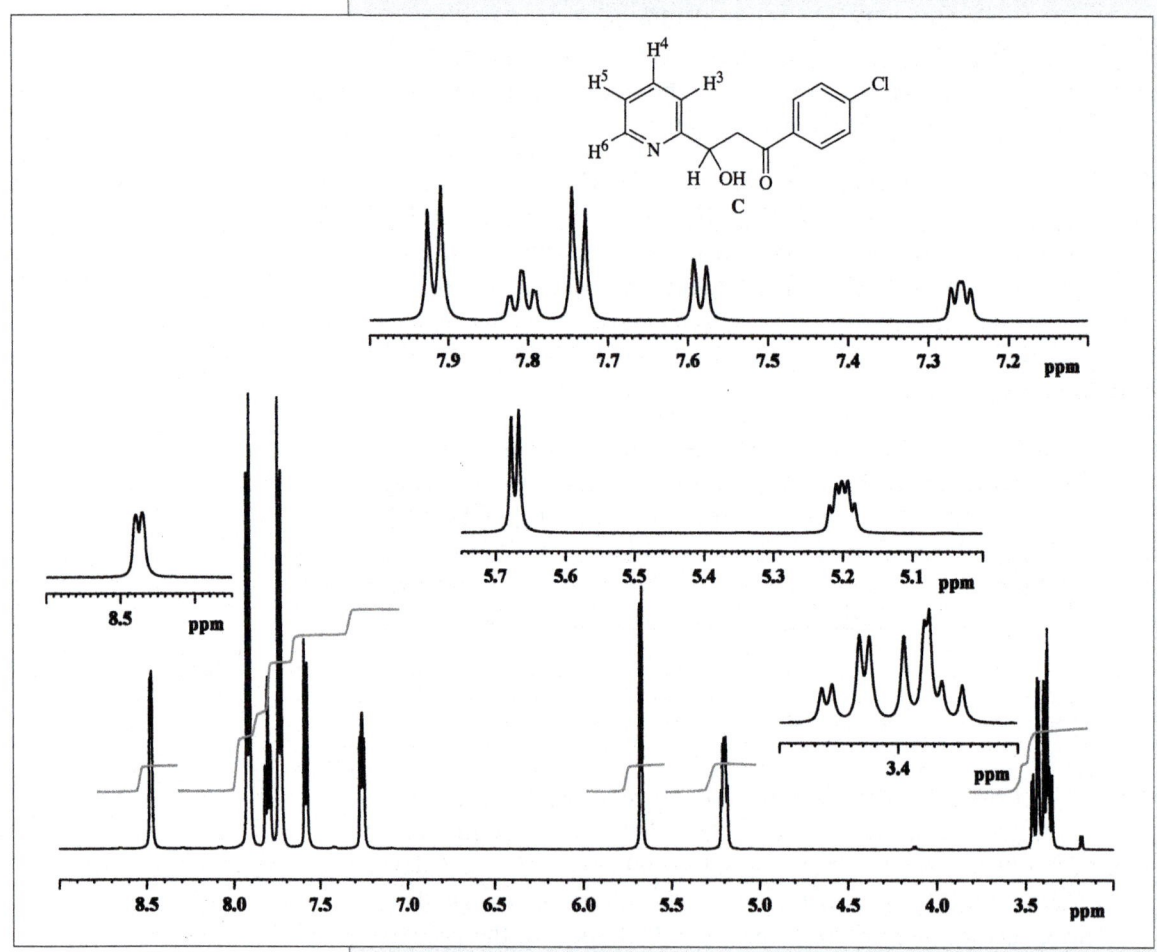

Figure 4.44   500 MHz ^{1}H NMR spectrum of a β-hydroxy ketone derivative **C** in DMSO-d_6

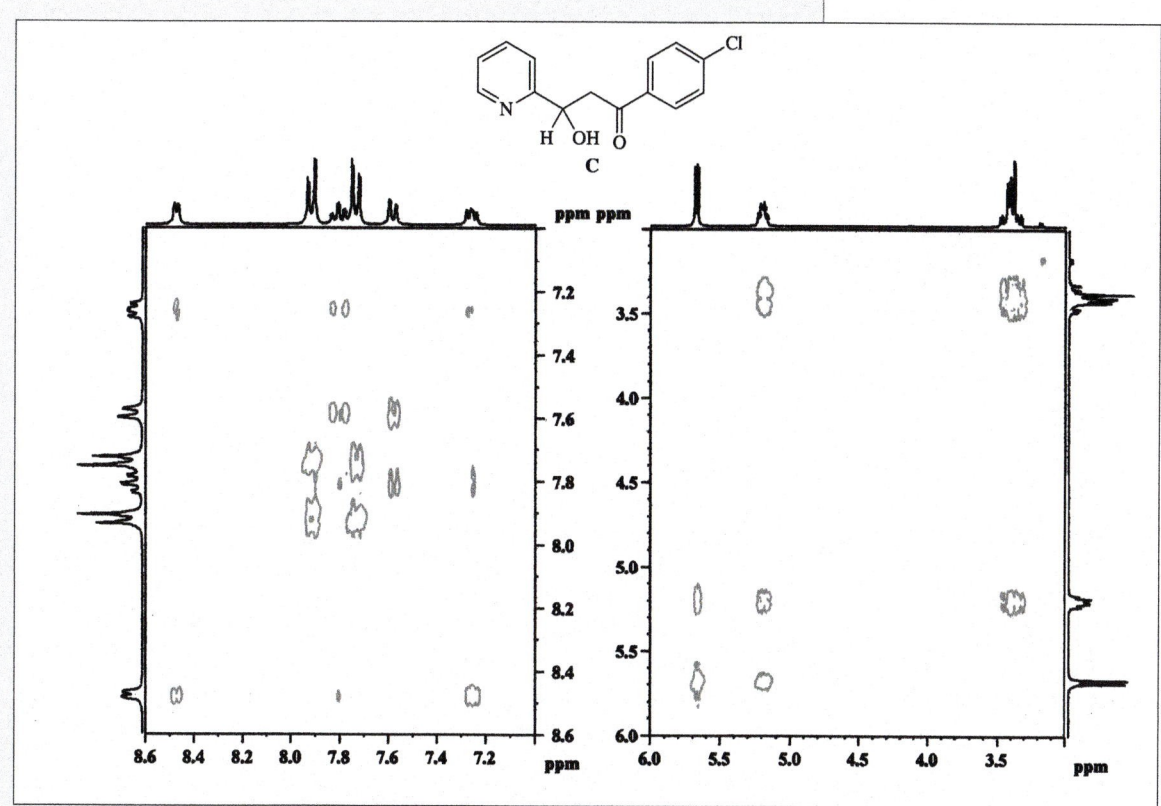

4.3. The oxidation of 1,4-dihydroxy-2-methylnaphthalene [δ_H (300 MHz, DMSO-d_6) 2.27 (3H, s), 6.62 (1H, s), 7.31–7.43 (2H, m), 7.98–8.08 (2H, m), 8.25 (1H, s), 9.36 (1H, s)] to the corresponding quinone, by O_2 in the presence of a phthalocyanine-iron(III) complex in dioxane (Figure 4.46), gave a product **D** with the ^{1}H NMR spectrum shown in Figure 4.47. Use this spectrum to determine whether the reaction has been successful and indicate the reasoning behind your deduction.

Figure 4.46 Oxidation of 1,4-dihydroxy-2-methylnaphthalene

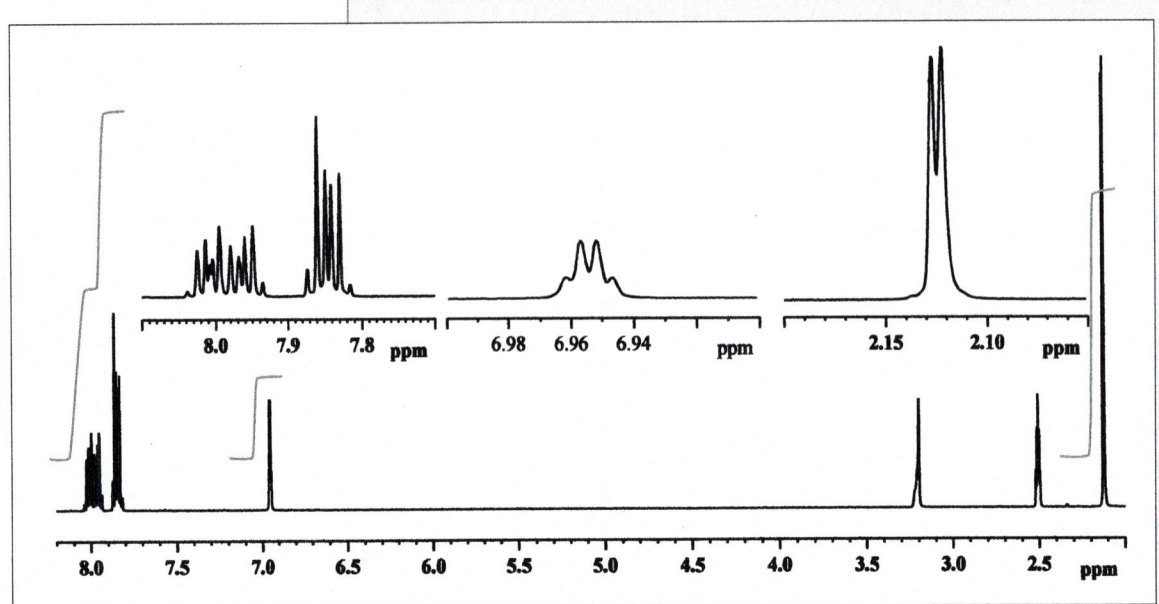

Figure 4.47   300 MHz ^{1}H NMR spectrum of oxidation product **D** in DMSO-d_6

4.4.   Assuming that $J_{AB} = 13.8\,\text{Hz}$, $J_{AX} = 9.6\,\text{Hz}$, $J_{BX} = 4.9\,\text{Hz}$ and $J_{XY} = 7.7\,\text{Hz}$, draw a diagram to scale to show the splitting patterns for the protons highlighted in colour in N-acetylphenylala-nine (N-ethanoylphenylalanine, **E**). Compare your predictions with the actual spectrum (Figure 4.20).

4.5.   The 300 MHz ^{1}H NMR spectrum of allyl hexanoate in CDCl$_3$ is shown in Figure 4.48. Use this spectrum and the data provided to calculate the coupling constants in the allyl ester portion of the molecule (CH$_2$=CH–CH$_2$OCOR).

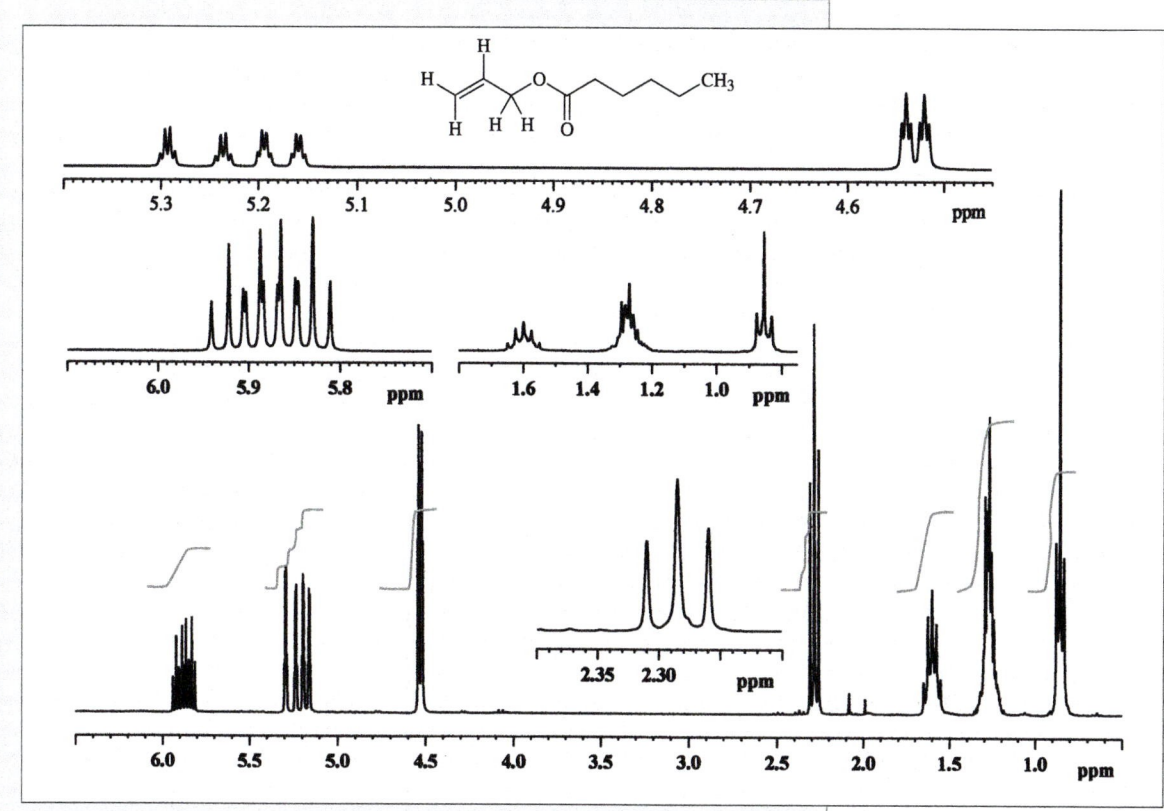

Figure 4.48 300 MHz ^{1}H NMR spectrum of allyl hexanoate in CDCl$_3$. The signals at δ 5.26 and 5.17 are both doublets of quartets, whilst that at δ 4.53 is a doublet of triplets

Frequency (Hz)	δ	Intensity	Frequency (Hz)	δ	Intensity
1590.950	5.3009	1.53	1556.869	5.1873	1.70
1589.639	5.2965	4.08	1550.428	5.1659	1.62
1588.292	5.2920	4.27	1549.123	5.1615	3.70
1586.969	5.2876	1.75	1547.799	5.1571	3.60
1573.739	5.2435	1.28	1546.467	5.1527	1.55
1572.426	5.2391	3.43	1363.650	4.5435	5.59
1571.090	5.2347	3.58	1362.324	4.5391	8.93
1569.773	5.2303	1.45	1360.972	4.5346	5.60
1560.837	5.2005	1.78	1357.967	4.5246	5.64
1559.523	5.1962	4.09	1356.625	4.5201	8.75
1558.211	5.1918	3.93	1355.264	4.5156	5.37

4.6. Use the 500 MHz ^{1}H (Figure 4.49) and HH COSY (Figure 4.50) NMR spectra of methyl β-glucopyranoside to fully assign all of the proton signals in this molecule. Note that the spectrum was obtained in D_2O so that all of the hydroxyl groups have been exchanged and any coupling to them has been lost (*hint*: find a signal for which there is only one coupling partner and work your way around the ring from this).

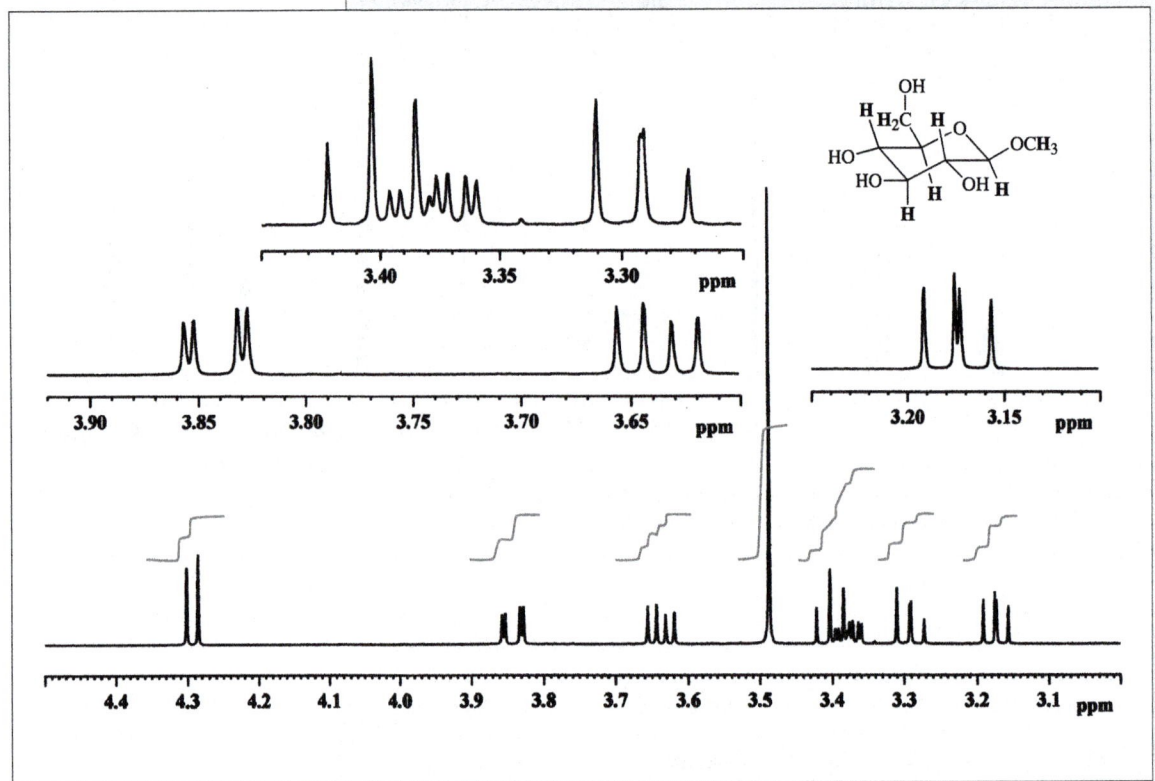

Figure 4.49 500 MHz ^{1}H NMR spectrum of methyl β-glucopyranoside in D_2O: $δ_H$ 3.17 (1H, dd, $J = 9.2$, 8.0 Hz), 3.29 (1H, dd, $J = 9.8$, 9.2 Hz), 3.37 (1H, ddd, $J = 9.8$, 6.0, 2.3 Hz), 3.40 (1H, t, $J = 9.2$ Hz), 3.49 (3H, s), 3.63 (1H, dd, $J = 12.3$, 6.0 Hz), 3.84 (1H, dd, $J = 12.3$, 2.3 Hz), 4.29 (1H, d, $J = 8.0$ Hz)

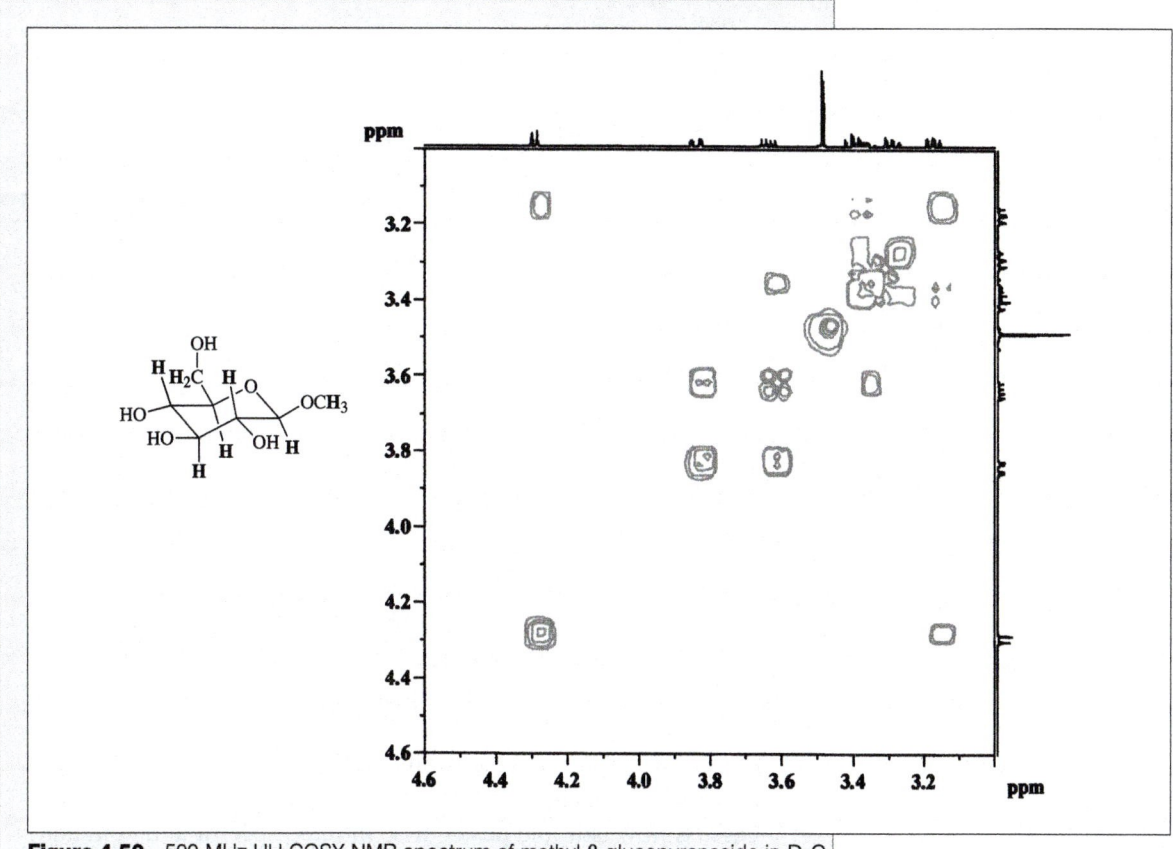

Figure 4.50　500 MHz HH COSY NMR spectrum of methyl β-glucopyranoside in D_2O

5
Mass Spectrometry

Aims

This chapter describes how mass spectrometry is used to determine the relative molecular mass of individual molecules and obtain information about their structure. After you have studied this chapter, you should be able to:

- Describe the processes involved in mass spectrometry
- Suggest an appropriate method for the vaporization and ionization of various analytes
- Explain how common fragments are formed from organic molecules
- Deduce structural information from mass spectra
- Describe the essential features of some methods of mass analysis
- Discuss how isotopic peaks arise in mass spectra and comment on their relative abundance
- Derive data on mass accuracy from high-resolution mass spectra
- Describe selected "hyphenated" mass spectrometry methods for particular applications

5.1 Instrumentation

Mass spectrometry is the technique most commonly used to measure the mass of molecules (usually organic molecules, including biomolecules), and can therefore help to characterize a particular molecule or to identify an unknown. Mass spectrometry can also be used to generate information about the structure of a molecule.

In order to generate a mass spectrum, which is a plot of intensity against mass-to-charge ratio (m/z), the sample must first be vaporized and ionized, then sorted by mass-to-charge ratio and finally detected. We can, therefore, break the process of mass spectrometry down into three stages: ion generation, mass analysis and ion detection.

5.1.1 Ion Generation

An ion source, as the name suggests, is the part of the mass spectrometer where the "gas phase ions" are generated. All of the various ion generation techniques that are available to us have one feature in common: they both vaporize and ionize our analyte molecules. In most cases, the vaporization stage occurs first, but with some techniques the ionization stage occurs first. The most commonly employed techniques are discussed in Section 5.2.

5.1.2 Mass Analysis

Once ionized, the analyte ions are separated by their interaction with an electric or magnetic field in a high vacuum (usually $\sim 10^{-4}$–10^{-7} N m^{-2}, which is 10^{-9}–10^{-12} bar) in order to minimize the interaction of the gaseous analyte ions with molecules in the air. In some cases, the mass analysis process can be made to produce data with high mass accuracy. The various options for the process of mass analysis are discussed in Section 5.4.

5.1.3 Ion Detection

The final part of the mass spectrometer is the ion detector, for which there are many options; however, discussion of the technical details of the various ion detection methods is beyond the scope of this text.

5.2 Vaporization and Ionization Processes

There are a number of vaporization and ionization processes that can be employed, but we will deal only with the most common techniques. In most cases the vaporization process occurs before the ionization takes place; however, there is one notable exception to this: electrospray ionization.

The techniques we will discuss are electron impact (EI) and chemical ionization (CI), fast atom bombardment (FAB), matrix assisted laser desorption ionization (MALDI) and electrospray ionization (ESI). Other techniques, such as field desorption (FD) and secondary ion mass

Historically, electron impact (EI) has been the ionization method most frequently employed, but this is a relatively harsh technique and has now been overtaken by electrospray ionization (ESI) in terms of everyday use.

spectrometry (SIMS), have their uses, but will not be discussed here, owing to our limit on space.

5.2.1 Ionization Techniques

Why would we choose one ionization technique over another? Well, each technique has a range of substrate (analyte) types and relative molecular masses for which it is best suited (as illustrated in Figure 5.1).

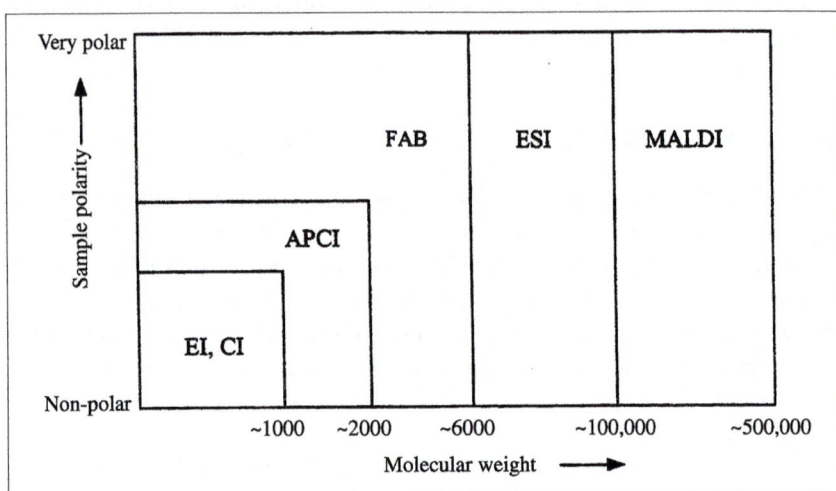

Figure 5.1 Applications for common ionization techniques

Electron Impact

Electron impact (EI) is a relatively harsh technique and involves a sample being volatilized into the gas phase by heating in a vacuum, then bombarded by a stream of electrons in order to cause ionization of the sample. These electrons are generated by a metallic filament and accelerated through a potential difference such that they have a typical energy of 70 eV (6.75×10^3 kJ mol^{-1}), and their impact on the gaseous sample molecules results in the ionization of these molecules, as shown in Figure 5.2. Now the first thing we might expect is that negatively charged electrons will lead to a negatively charged ion (an anion). This is not the case, as the electrons are moving too rapidly to be captured by the molecule (the molecule would also have to contain groups capable of capturing an electron), and when they impact with the molecule they actually knock an electron out of the molecule (Figure 5.3), resulting in a cation radical (a positively charged ion with an unpaired electron).

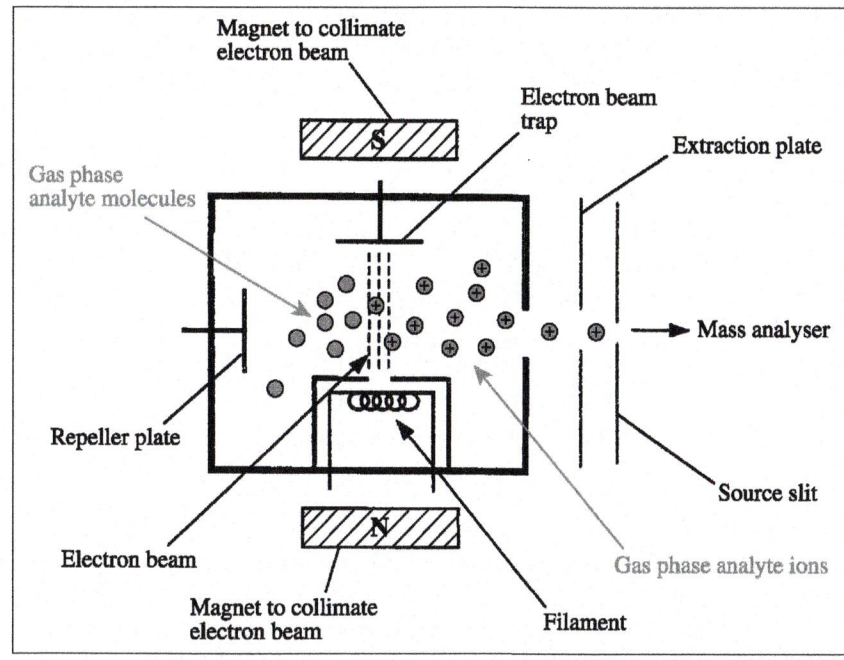

Figure 5.2 Diagram of an EI source

The electron lost will be one of the least tightly bound in the molecule, *i.e.* one of the electrons in the highest occupied molecular orbital (HOMO), and, in general, the order of ease with which electrons are lost upon EI is:

$$M + e^- \longrightarrow M^{+\bullet} + 2e^-$$

Figure 5.3 EI process

$$\text{lone pair} > \pi\text{-bonded pair} > \sigma\text{-bonded pair}$$

We said earlier that EI is a relatively harsh technique and we will now see why. The amount of energy required to remove an electron from a molecule (which depends upon what type of orbital the electron occupies) is approximately 7 eV (675 kJ mol^{-1}), so that the electrons employed in EI have ten times the energy required to do the job. Some of this excess energy is imparted to the molecule and results in an excess of vibrational energy and the fragmentation (breaking up) of the molecular ion (see Section 5.3). In some cases, the extent of fragmentation results in the absence of the molecular ion.

Chemical Ionization

Chemical ionization (CI) is closely related to electron impact, as it also uses a stream of electrons in the ionization process. In this case, however, it is not the sample molecules which are ionized, but a reagent gas, usually ammonia or methane, which is present at a much higher concentration.

Once again, the sample must be volatilized by heating in a vacuum, but the main difference between CI and EI is that in CI the sample is ionized by a strong acid produced by the ionization of the reagent gas (Scheme 5.1).

$$
\begin{array}{lll}
\text{(a)} & CH_4 & \xrightarrow{\text{electron impact}} CH_4^{+\bullet} + 2e^- \\
\text{(b)} & CH_4^{+\bullet} + CH_4 & \longrightarrow \ {}^+CH_5 + {}^\bullet CH_3 \\
\text{(c)} & {}^+CH_5 + M & \longrightarrow \ {}^+MH + CH_4
\end{array}
$$

$$
\begin{array}{ll}
NH_3 & \xrightarrow{\text{electron impact}} NH_3^{+\bullet} + 2e^- \\
NH_3^{+\bullet} + NH_3 & \longrightarrow \ {}^+NH_4 + {}^\bullet NH_2 \\
{}^+NH_4 + M & \longrightarrow \ {}^+MH + NH_3
\end{array}
$$

Scheme 5.1 CI processes

As we can see in Scheme 5.1, the electron impact on the reagent gas (methane or ammonia) leads to a molecular ion, $CH_4^{+\bullet}$ (Scheme 5.1a), which reacts with the CH_4 reagent gas to give a strong acid, CH_5^+, and a radical, $CH_3^\bullet$ (Scheme 5.1b). It is this strong acid (CH_5^+) which ionizes the sample by protonation (Scheme 5.1c).

As mentioned above, the two most commonly used reagent gases are ammonia (NH_3) and methane (CH_4). Chemical ionization generally results in the production of an $[M+1]^+$ ion with little excess energy, and so fragmentation is less evident than in EI. The acid formed from ammonia (NH_4^+) is not as strong as that from methane (CH_5^+), so that, in cases where the sample is ionized by the ammonium ion (NH_4^+), this process is less energetically favourable than the corresponding process with the carbonium ion (CH_5^+) and, therefore, less fragmentation usually accompanies the use of ammonia as the reagent gas.

Chemical ionization, unlike electron impact, is also capable of producing negatively charged ions. EI does not produce anions because the kinetic energy (and so the velocity) of the electrons is too great for them to be captured by a molecule. We can liken this process to trying to catch a tennis serve whilst cycling – a very difficult (some might say, impossible) task. If, however, the ball is moving at a much slower speed, such as, for example, a tennis ball thrown to a cyclist, then the task becomes easier. In chemical ionization, owing to the presence of relatively high concentrations of the reagent gas, collisions are much more likely between electrons and the reagent gas, thereby reducing the speed of the electrons and making them more likely to be captured by the molecule, so giving an anion (Scheme 5.2). Alternatively, a reagent gas anion, *e.g.* X^-, may be produced in the source, and this can remove a proton from the sample, also leading to the formation of an anion (Scheme 5.2).

The carbonium ion, CH_5^+, being a strong acid, will usually protonate a sample molecule, but it is often the case that the ammonium ion, being a weaker acid, will not protonate a sample unless it contains very basic functional groups.

$$M + e^- \longrightarrow M^{-\bullet} \longrightarrow A^{\bullet} + :B^-$$
$$M-H + X^{\bar{\bullet}} \longrightarrow M^{\bar{\bullet}} + X-H$$

Scheme 5.2 Formation of anions by CI

Fast Atom Bombardment

We will now concentrate upon some "soft" ionization methods – so called because they give rise mainly to the peak associated with the molecular ion and very little fragmentation. Another benefit of these techniques is that, unlike both EI and CI, they do not require the sample to be volatile and so permit the analysis of biomolecules, which are generally large, sensitive and polar.

Mass spectrometry has been revolutionized by the advent of electrospray ionization, but, before we concentrate on this relatively recent addition to the array of ionization methods, we will first discuss two other techniques which are routinely used for the ionization of biomolecules: matrix assisted laser desorption ionization (MALDI) and fast atom bombardment (FAB). These techniques share common features in that:

- The analyte (sample being analysed) is dissolved in a low-freezing-point matrix, which not only keeps the analyte in solution in the high vacuum ion source, but also assists in the vaporization and ionization processes.
- The solution is given a large pulse of energy, either from a beam of fast moving atoms (FAB) or from a laser (MALDI).

Fast atom bombardment (FAB), as the name implies, involves bombarding a solution of the analyte in a matrix (most usually propane-1,2,3-triol, propane-1,2,3-trithiol, 2-nitrobenzyl alcohol or triethanolamine, Figure 5.4) with a beam of fast moving atoms, generally xenon atoms with energy in the range 6–9 keV (580–870 kJ mol^{-1}).

The matrix helps to "protect" the analyte as, if it were absent, the direct bombardment of the analyte by the fast atoms would lead to extensive fragmentation.

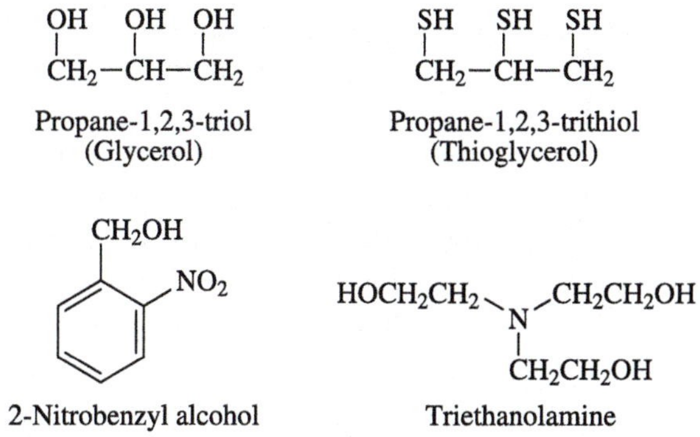

Figure 5.4 Structures of FAB matrices

This bombardment results in the transfer of energy from the Xe atoms to the matrix, leading to the breaking of intermolecular bonds and the desorption of the analyte (usually as an ion) into the gas phase. Unlike EI, FAB can also be used to generate negatively charged ions.

FAB has been widely used for the ionization of large polar molecules and generally gives $[M + 1]^+$ peaks (corresponding to the MH^+ ion) with little fragmentation. In negative ion mode the most abundant peaks obtained are the $[M - 1]^-$ peaks, corresponding to $[M - H]^-$. A feature of FAB spectra is the peaks which correspond to protonated (or deprotonated) clusters of the matrix. For example, if propane-1,2,3-triol is used as the matrix, then peaks would be obtained for its protonated oligomers, $[(HOCH_2CHOHCH_2OH)_nH]^+$, at m/z values of 93 (for $n = 1$), 185 (for $n = 2$), *etc.* In addition, some minor, and useful, fragmentation is sometimes observed as a result of FAB ionization.

Matrix Assisted Laser Desorption Ionization

As stated earlier, matrix assisted laser desorption ionization (MALDI) is similar in principle to FAB except that in this case the energy is transferred to the matrix from a laser beam and the matrix employed must therefore have a chromophore which absorbs at the wavelength of the laser. Common matrices employed in MALDI are aromatic or heterocyclic carboxylic acids, such as 2,5-dihydroxybenzoic acid and nicotinic acid (Figure 5.5). The matrix absorbs a pulse of energy from the laser beam and undergoes rapid heating, which ultimately leads to the vaporization and ionization of the analyte molecules. Once again, peaks corresponding to matrix cluster ions are obtained along with the usual MH^+ peaks for the molecule under investigation.

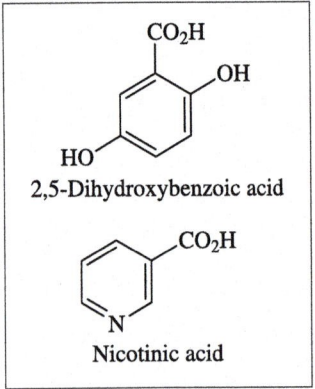

2,5-Dihydroxybenzoic acid

Nicotinic acid

Figure 5.5   Common MALDI matrices

Electrospray Ionization

Electrospray ionization (ESI) was first employed more than 20 years ago, but it is fairly recently that it became a routine technique for the "soft" ionization of a wide range of polar analytes, including biomolecules. For this technique, the analyte is usually dissolved in a mixture of an organic solvent (most commonly acetonitrile or methanol) and water with a pH modifier [*e.g.* formic (methanoic) or acetic (ethanoic) acid for positive ion mode]. The presence of the pH modifier ensures that ionization takes place in the solution state. This is the only common case where ionization occurs before ion vaporization; the exact mechanism of the vaporization (Figure 5.6) is still not clearly understood in ESI.

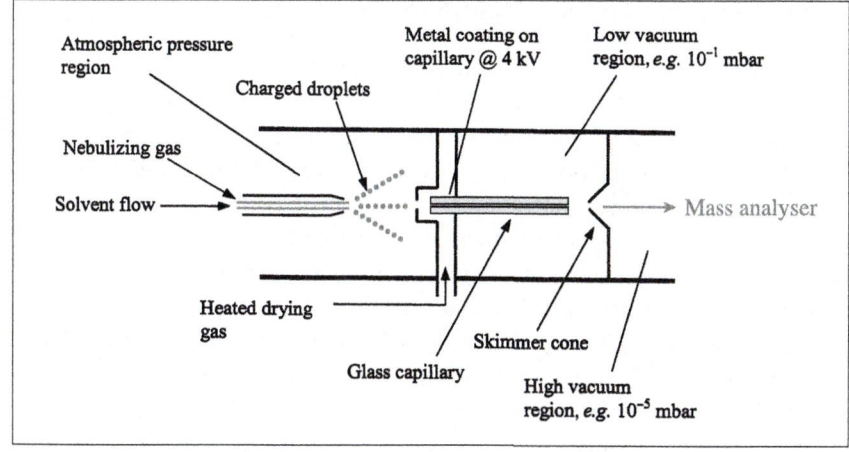

Figure 5.6 Schematic diagram of an electrospray source

Because ionization has taken place in the solution state by protonation or deprotonation of the analyte (depending upon the pH modifier used), the molecular species detected is almost exclusively $[M+H]^+$ in positive ion mode and $[M-H]^-$ in negative ion mode, and both these species undergo very little fragmentation. Since the solvent mixtures employed in ESI are those commonly used in reverse phase liquid chromatography, ESI is frequently combined with LC to give liquid chromatography–mass spectrometry (LC-MS) analysis (see Section 5.6). One further advantage of ESI is that it often gives multiply charged ions for large molecules with many ionizable functional groups. This has the advantage of lowering the m/z ratio and thereby allowing the determination of the masses of large molecules without the need for a detector that has a large mass range. One disadvantage of ESI is that it is very sensitive to contaminants in the solvents, particularly alkali metals, and we often see ions which correspond to $[M+Na]^+$ or $[M+NH_4]^+$. These peaks, however, can often be useful in accurate mass determination (see Section 5.5) using ESI.

The singly charged peak for an ion of relative molecular mass 10,000 would appear at m/z 10,000, while the doubly charged ion, M^{2+}, would appear at m/z 5000.

Atmospheric Pressure Chemical Ionization

The advent of atmospheric pressure chemical ionization (APCI) is a relatively recent development, in which the same processes occur as in CI, outlined previously, but at atmospheric pressure. By a very similar mechanism to CI, the reagent gas (water) becomes protonated and can act as an acid towards the analyte, leading to the addition of a proton. Once again the species formed in positive ion mode is $[M+H]^+$. In the case of negative ion mode, the reagent gas acts as a base towards the analyte, and deprotonation occurs leading to the formation of $[M-H]^-$. Once ions have been formed, they are funnelled towards the analyser inlet of the MS instrument by the use of electric potentials. APCI is also employed in LC-MS systems (see Section 5.6).

5.3   Fragmentation Processes

As we mentioned earlier, the main aim of mass spectrometry is to obtain a peak for the molecular ion, $M^{+\bullet}$, since this can be used to confirm, or to obtain the relative molecular mass. Using EI or CI it is often the case that we obtain peaks for additional ions with smaller masses (or, more correctly, smaller mass to charge ratio, m/z), produced by the fragmentation of the molecular ion, and that these fragment ions dominate the mass spectrum. As discussed in Section 5.2.1, the nature of these ionization processes means that the molecular ion is often produced in an excited state, with excess vibrational energy, and this leads to fragmentation of the weakest bonds in the molecule. In general, there is a lower degree of fragmentation when using CI than with EI, while ESI and FAB are comparatively "soft" ionization techniques and generally give rise to molecular ions only.

The cation radical of the molecular ion can fragment to give daughter ions *via* the loss of either a radical or a neutral molecule (Scheme 5.3). The process does not have to stop here, and both B^+ and $C^{+\bullet}$ can also fragment further, so that the peaks for ions with even smaller masses can arise from the fragmentation of either the parent or daughter ion.

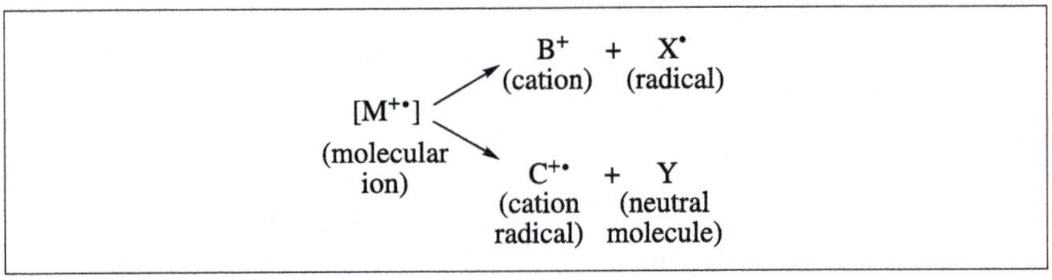

Scheme 5.3 Fragmentation of the molecular cation radical

We will now look at some of the most common fragmentation processes, starting with the fragmentation of alkanes, *e.g.* hexane, C_6H_{14} (Scheme 5.4).

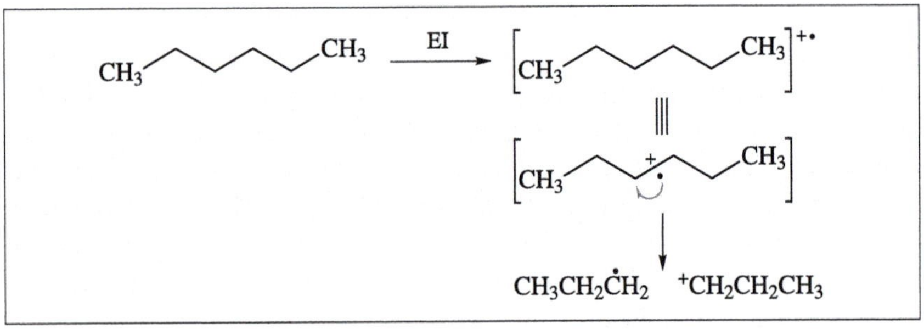

Scheme 5.4 The fragmentation of hexane

For an alkane, the ionization process will involve the loss of an electron from a σ-bond, as shown in Scheme 5.4 for the loss of an electron from the central C–C bond of hexane (the loss of an electron from all of the σ-bonds in this molecule is almost equally probable, so a large number of fragments can be produced; Scheme 5.5).

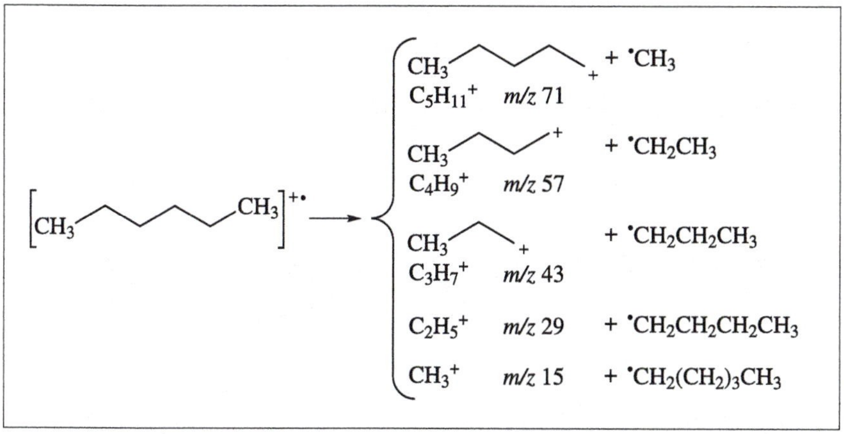

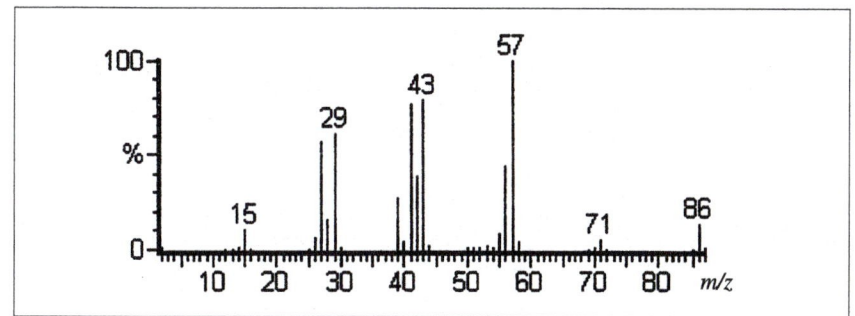

Scheme 5.5 Alternative fragmentation pathways for hexane

In the EI MS of hexane, then, we would expect to see the molecular ion, m/z 86, as well as ions due to the fragmentation of each bond (Figure 5.7).

Figure 5.7 EI MS of hexane

This series of daughter ions differing in their mass by 14 is characteristic of the straight-chain alkanes (homologous series differing by a CH_2 group).

Fragmentations that give rise to stable carbocations will be particularly favoured (Box 5.1). The order of carbocation stability is shown in Figure 5.8 and is related to the reduction in the positive

An **homologous series** is a family of related compounds with the same general formula that differ by a constant unit between each member and the next. For example, in the series of alkanes of general formula C_nH_{2n+2}, each member differs from the next by a CH_2 unit.

charge on any carbon atom through electron donation (with the more electron-donating alkyl groups, the smaller the positive charge) or delocalization through resonance (in which case the positive charge is "spread out" over the molecule, with the more resonance forms the better).

Box 5.1 Fragmentation of Branched Chain Alkanes

In branched chain alkanes, fragmentation often occurs next to the branch site, allowing formation of the more stable secondary, or even tertiary, carbocation:

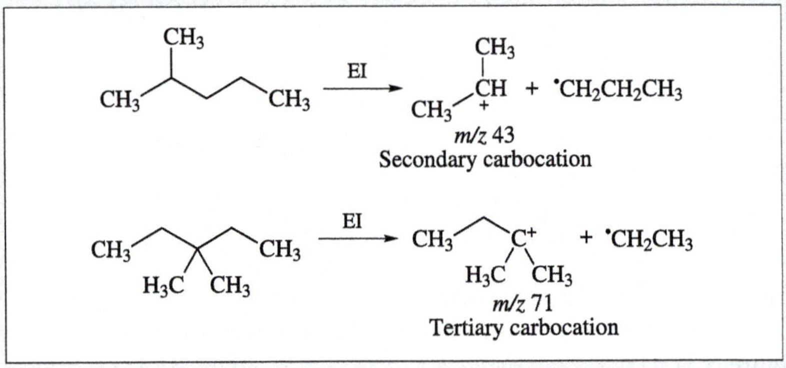

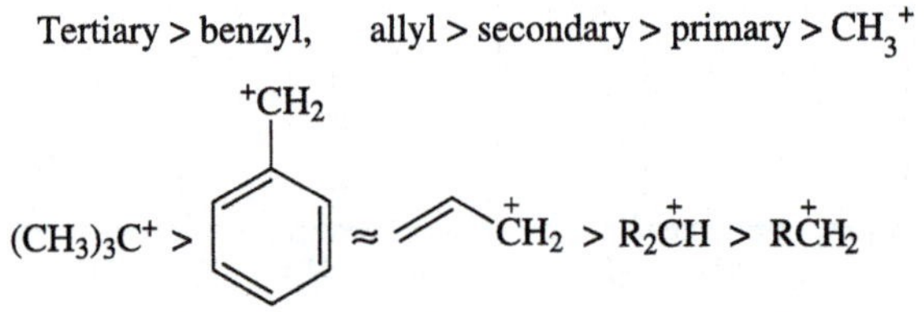

Figure 5.8 Order of carbocation stability

One particularly stable carbocation is the tropylium ion. The tropylium ion is formed by loss of a leaving group from 7-substituted cyclohepta-1,3,5-trienes. The carbocation formed is cyclic, planar and has six π-electrons in the bonding π molecular orbitals, *i.e.* it is aromatic. In fact, this ion is so stable that a hydride ion can be lost from cyclohepta-1,3,5-triene with relative ease (Scheme 5.6).

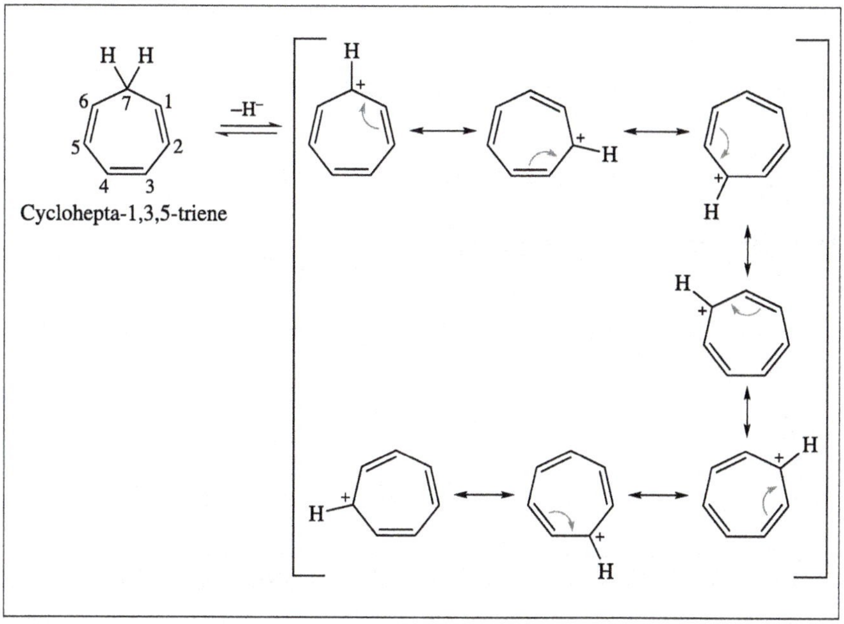

Scheme 5.6 Resonance stabilization of the tropylium ion

You will not be surprised, then, to learn that the tropylium ion is the major peak found in the MS of 7-substituted cyclohepta-1,3,5-trienes, but you may be surprised to find that the benzyl carbocation rearranges to the tropylium ion under MS conditions (Scheme 5.7).

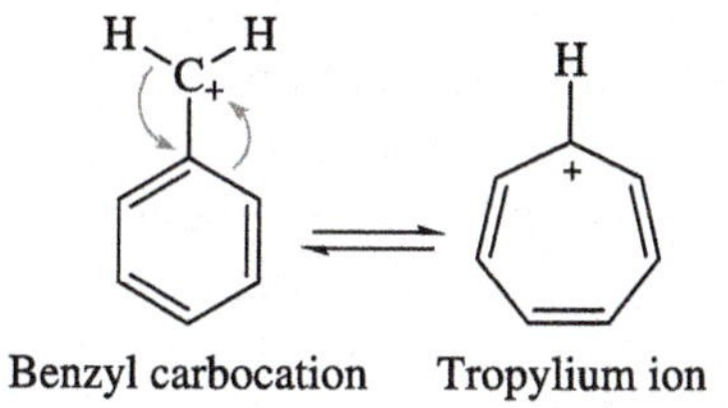

Benzyl carbocation   Tropylium ion

Scheme 5.7 Rearrangement of the benzyl cation to the tropylium ion

How we can be sure that this is happening, when both ions appear in the MS at m/z 91? This debate can be settled using carbon (^{13}C) labelling of the benzylic CH_2 atom. If the benzyl cation remains unaltered, subsequent fragmentation will lead to the loss of a labelled $^{13}CH_2{}^+$ unit of m/z 15 (13 + 2), but if the benzyl cation undergoes rearrangement to give the tropylium ion, the labelled carbon will become identical to the other carbon atoms in the ring and there will be a 1 in 7 chance of finding a labelled $^{13}CH^+$ unit of m/z 14 (13 + 1) during subsequent fragmentations.

A qualitative judgement can be made by comparing the MS of the tropylium and benzyl cations, generated from tropylium hexafluorophosphate and benzyl bromide, respectively, as shown in Figure 5.9.

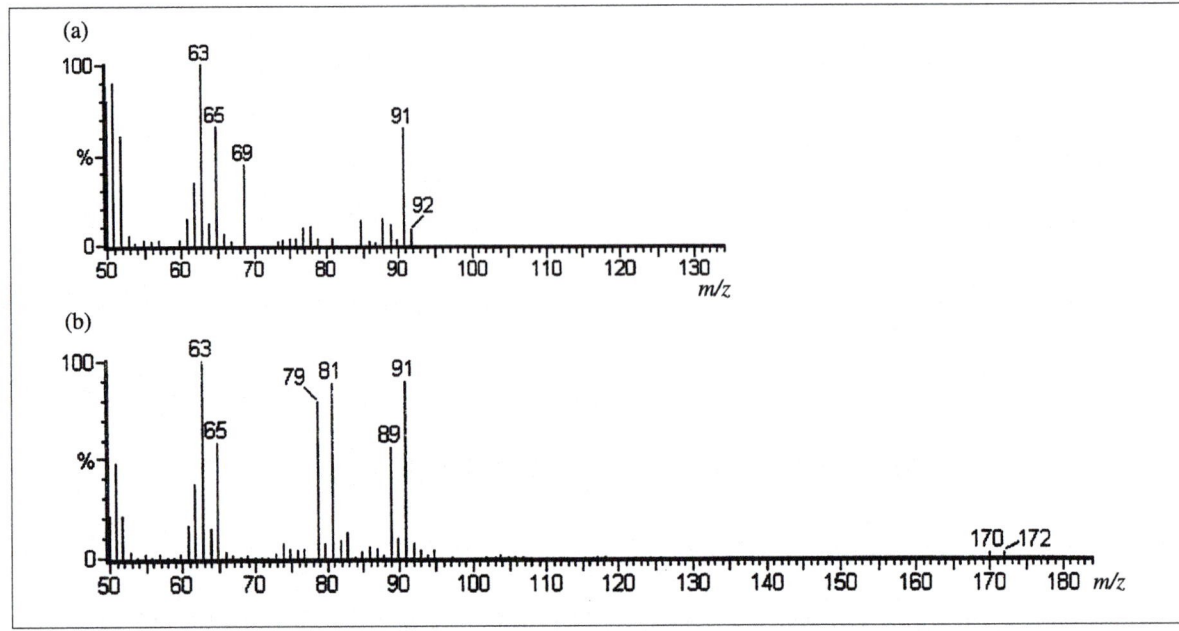

Figure 5.9   The EI mass spectra of (a) tropylium hexafluorophosphate and (b) benzyl bromide

As we can see, the MS fragmentation patterns are very similar and support the common cation theory [note that the ions at m/z 79 and 81 in Figure 5.9(b) are probably due to $Br^{+\bullet}$; see Section 5.5.1].

Other stable carbocations are those with an adjacent heteroatom, *e.g.* oxygen, which can stabilize the cation through resonance (Scheme 5.8).

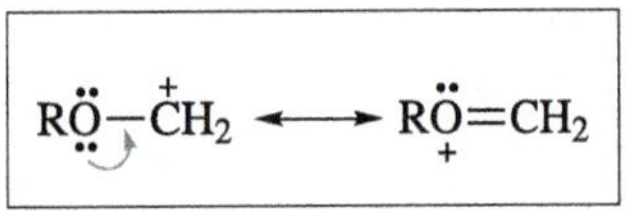

For molecules containing heteroatoms (O, N, Cl, Br, *etc.*), a very common fragmentation is the cleavage of the α,β-bond (often referred to as cleavage β to the heteroatom). In such molecules, the lone pairs on the heteroatom will be the least tightly bound in the molecule, and it will be one of these electrons which is lost upon electron impact (Scheme 5.9), leading to cleavage β to the heteroatom.

$$R\ddot{O}-\overset{+}{C}H_2 \longleftrightarrow R\ddot{O}=CH_2$$

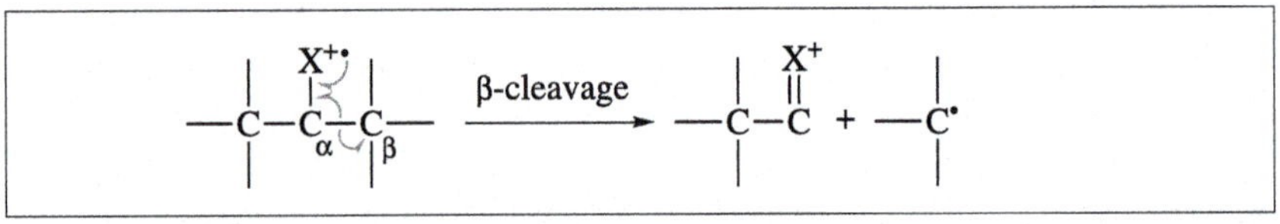

Scheme 5.9   Cleavage of the α,β-bond adjacent to a heteroatom

Some examples of cleavage β to a heteroatom are shown in the following figures. Figure 5.10 shows the EI MS of benzaldehyde (C_7H_6O), which, in common with other aldehydes, loses H^+ due to β-cleavage, so we see the $M^{+\bullet}$ peak at m/z 106 and the $[M-1]^+$ peak at m/z 105 as the major peaks. The benzoyl cation is particularly stable owing to extensive electron resonance (Scheme 5.10), a fact that is reflected by the intensity of the peak due to this fragment.

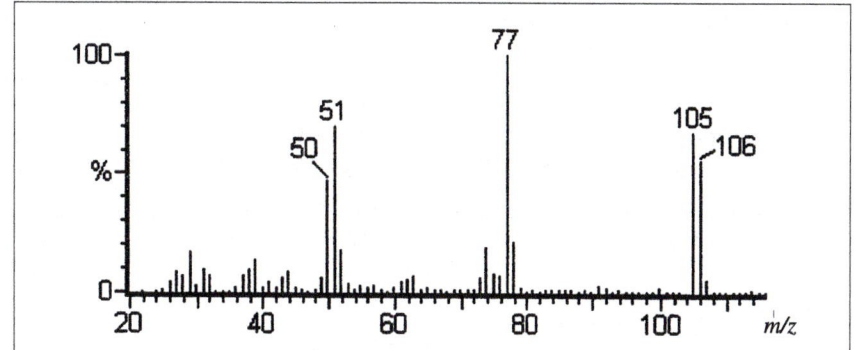

Figure 5.10   The EI MS of benzaldehyde

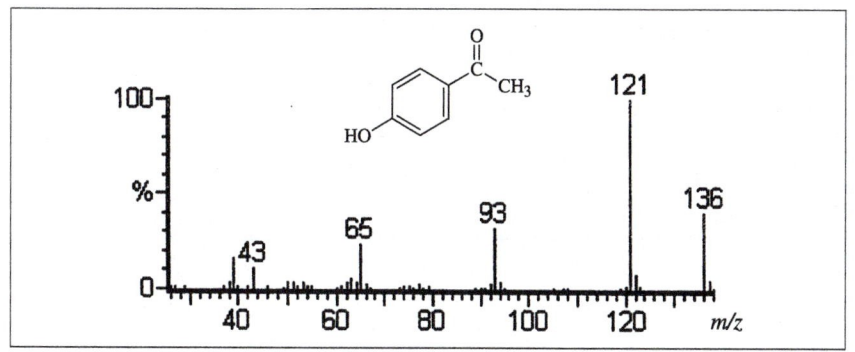

Scheme 5.10   Resonance stabilization of the benzoyl cation

Figure 5.11 shows the EI MS of 4'-hydroxyacetophenone ($C_8H_8O_2$), with the expected $M^{+\bullet}$ peak at m/z 136.

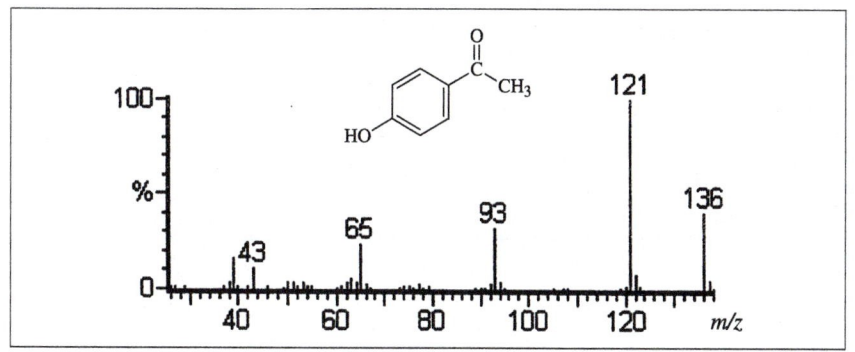

Figure 5.11   The EI MS of 4'-hydroxyacetophenone

Using Scheme 5.9 as a guide, draw the two β-cleavage pathways that are possible for this molecule and use these to explain the peaks at m/z 121 and 43.

This compound can undergo β-cleavage in two different ways: to lose either the methyl group or the 4-hydroxyphenyl group (Scheme 5.11), and we can see peaks for both of the possible β-cleavage products in Figure 5.11: $[M-CH_3]^+$ at m/z 121 and $[M-C_6H_4OH]^+$ at m/z 43.

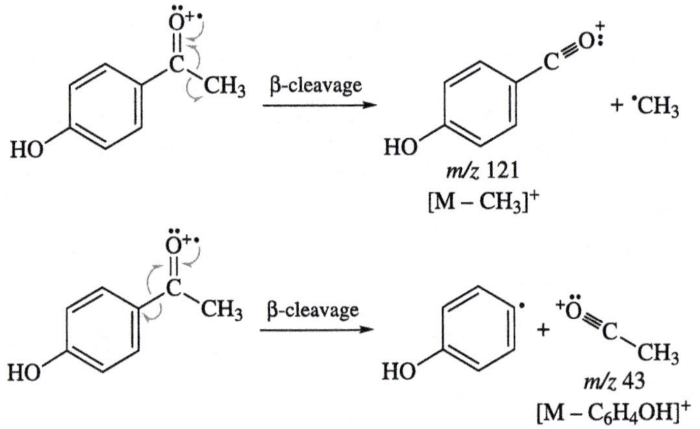

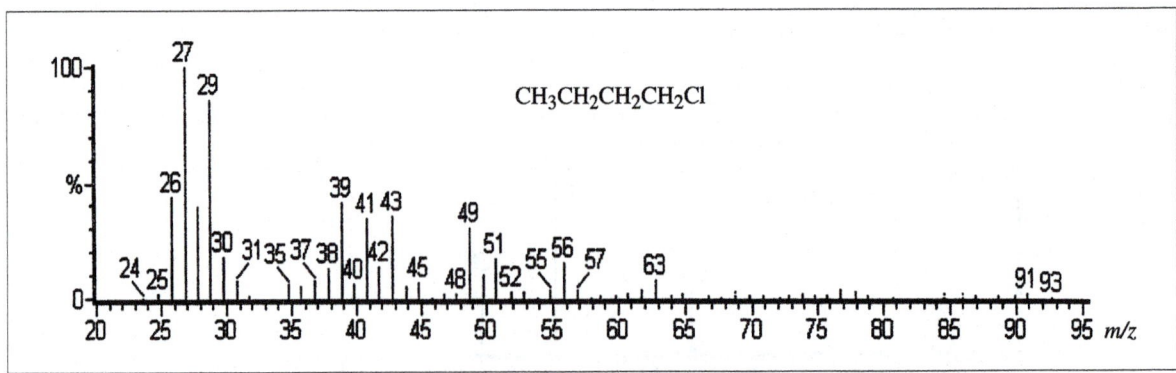

Scheme 5.11   The two β-cleavage pathways for 4′-hydroxyacetophenone

Figure 5.12 shows the EI mass spectrum of 1-chlorobutane, which we might expect to have a molecular ion, $M^{+\bullet}$, of m/z 94/92 and a fragment due to β-cleavage, $^+CH_2Cl$, at m/z 51/49 (two peaks due to chlorine's isotope pattern – see Section 5.5.1) (Scheme 5.12).

100─

%─

0─

27
29
26
24 25
30 31 35 37 38
39 41 43
40 42
49
45 48
51 52
55 56 57
63
91 93

20   25   30   35   40   45   50   55   60   65   70   75   80   85   90   95 m/z

$CH_3CH_2CH_2CH_2Cl$

Figure 5.12   The EI MS of 1-chlorobutane

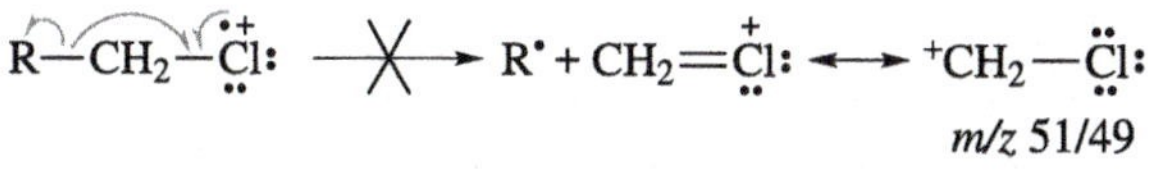

$$R-CH_2-\overset{\bullet+}{\underset{\bullet\bullet}{Cl}}: \quad \xrightarrow{\quad\times\quad} \quad R^{\bullet} + CH_2=\overset{+}{\underset{\bullet\bullet}{Cl}}: \longleftrightarrow {}^+CH_2-\overset{\bullet\bullet}{\underset{\bullet\bullet}{Cl}}:$$
$$m/z\ 51/49$$

Scheme 5.12 β-Cleavage of a chloroalkane

However, β-cleavage is not very common for haloalkanes, which tend not to form ions that have a positively charged halogen. Loss of halide radical is usually more common, and most haloalkanes give a major fragment of $[M-X]^{+\bullet}$; none of the expected β-cleavage peaks are observed. For haloalkanes such as 1-chlorobutane, which can form a cyclic ion, the major peak is due to loss of HCl by a two-step fragmentation, as shown in Scheme 5.13. The cyclic ion is also often observed in the mass spectrum (in this case formed by the loss of $H^{\bullet}$ to give the very small peaks at m/z 91 and 93).

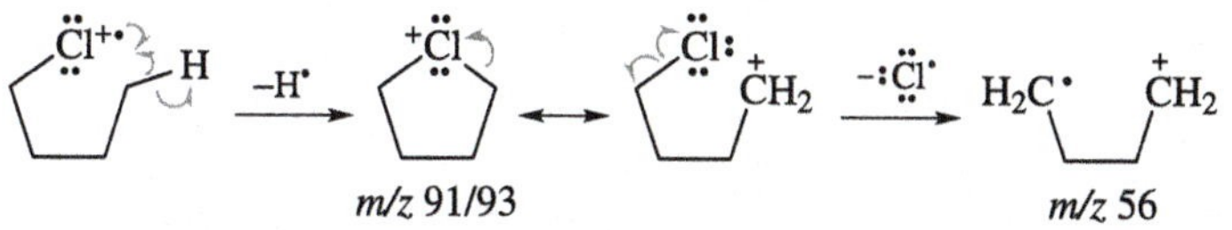

Scheme 5.13 Loss of hydrogen and chlorine radicals from 1-chlorobutane

You may have already come across the importance of six-membered ring transition states in organic chemistry, *e.g.* in the decarboxylation of β-keto acids (Scheme 5.14).

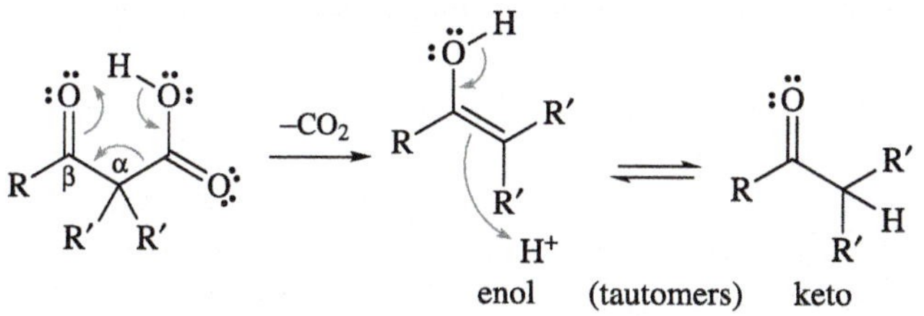

Scheme 5.14 The decarboxylation of β-keto acids

Two other important fragmentations in MS also involve cyclic six-membered transition states: (i) the McLafferty rearrangement and (ii) the retro-Diels–Alder reaction.

The McLafferty rearrangement is shown in Scheme 5.15. As an example, the peak at m/z 72 in the EI MS of heptan-3-one (representing a loss of 42 from the molecular ion, $M^{+\bullet}$, m/z 114) (Figure 5.13) is formed, which is common in carbonyl compounds such as esters and ketones that have a hydrogen atom on the γ-carbon to the carbonyl.

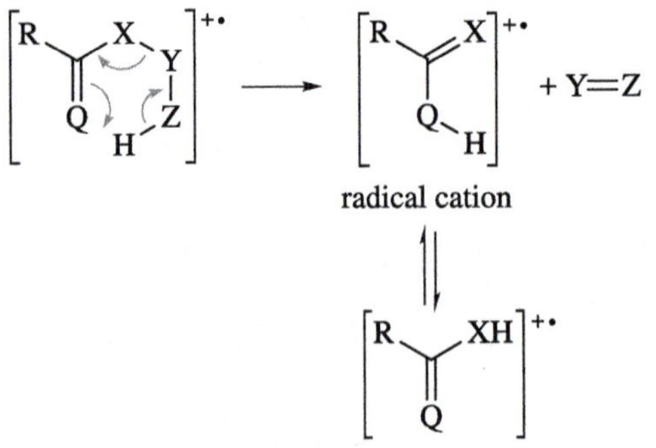

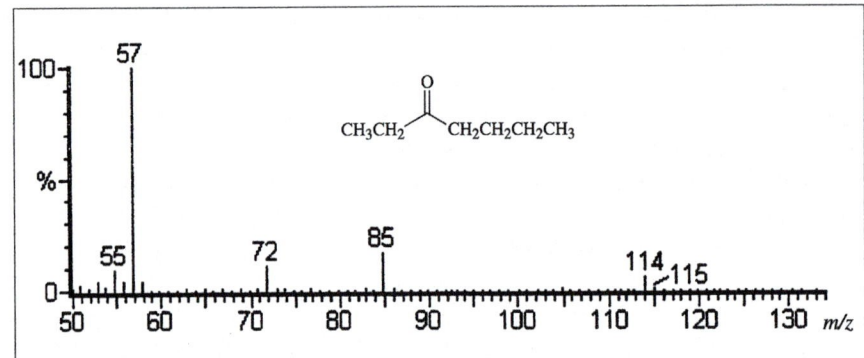

Scheme 5.15   The McLafferty rearrangement

Figure 5.13   The EI MS of heptan-3-one

Using Scheme 5.15 as a guide, draw the McLafferty rearrangement of heptan-3-one, then compare your answer to Scheme 5.16 below.

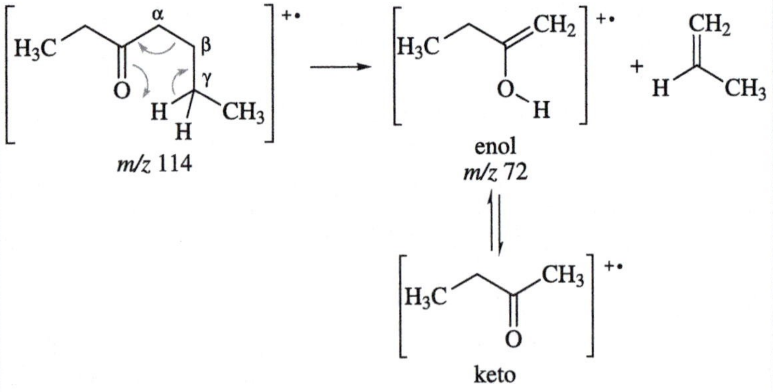

Scheme 5.16   The McLafferty rearrangement of heptan-3-one

The Diels–Alder reaction results in a $[4\pi + 2\pi]$ cycloaddition to give a six-membered ring. The reverse process, the retro-Diels–Alder fragmentation, results in ring opening, and is also common in the MS of six-membered rings containing a double bond (Scheme 5.17).

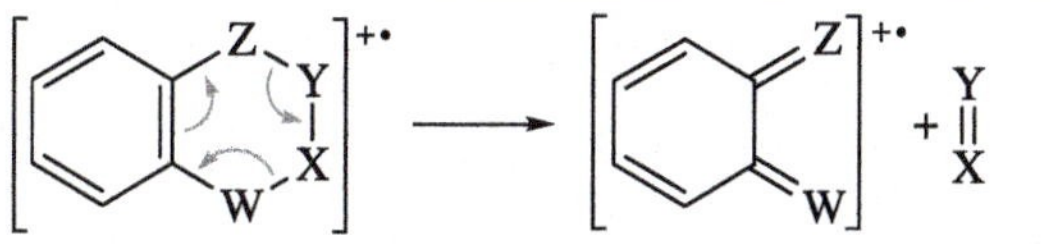

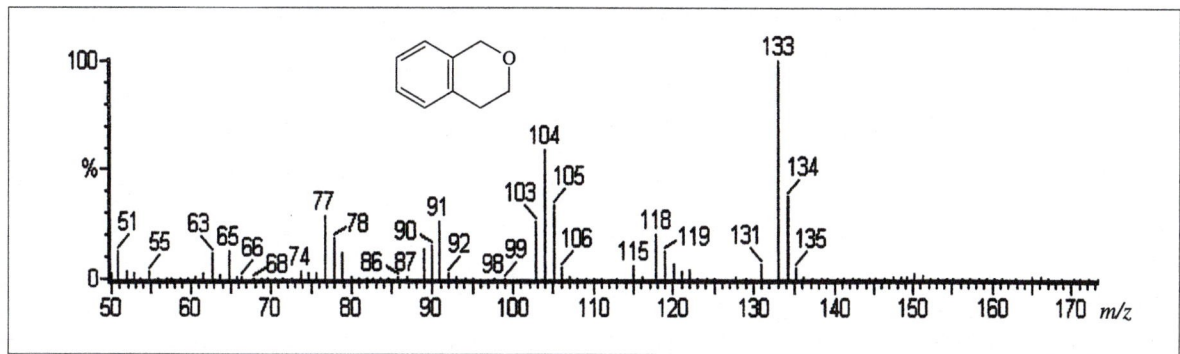

Scheme 5.17 The retro-Diels–Alder fragmentation

For example, the peak at m/z 104 in the EI MS of benzo[c]pyran ($M^{+\bullet}$ m/z 134, Figure 5.14), representing a loss of 30 mass units (CH_2O), is formed by the retro-Diels–Alder ring opening of the pyran ring (Scheme 5.18).

Figure 5.14 The EI MS of benzo[c]pyran (isochroman)

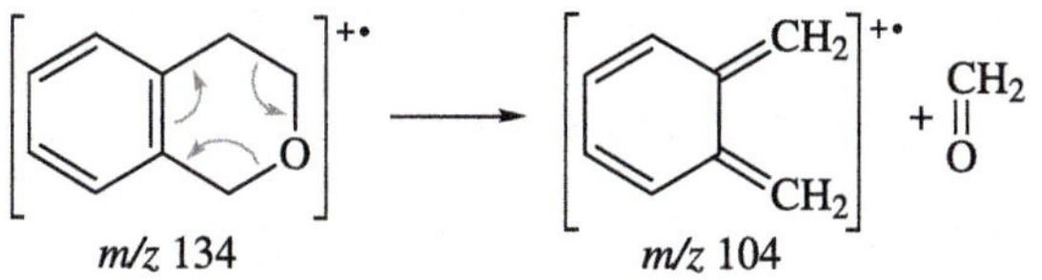

Scheme 5.18 The retro-Diels–Alder fragmentation of benzo[c]-pyran

We have already seen examples of β-cleavage in this section. Consider the MS of methyl 2-hydroxybenzoate ($M^{+\bullet}$ m/z 152, Figure 5.15), and explain how the fragment at m/z 120 arises.

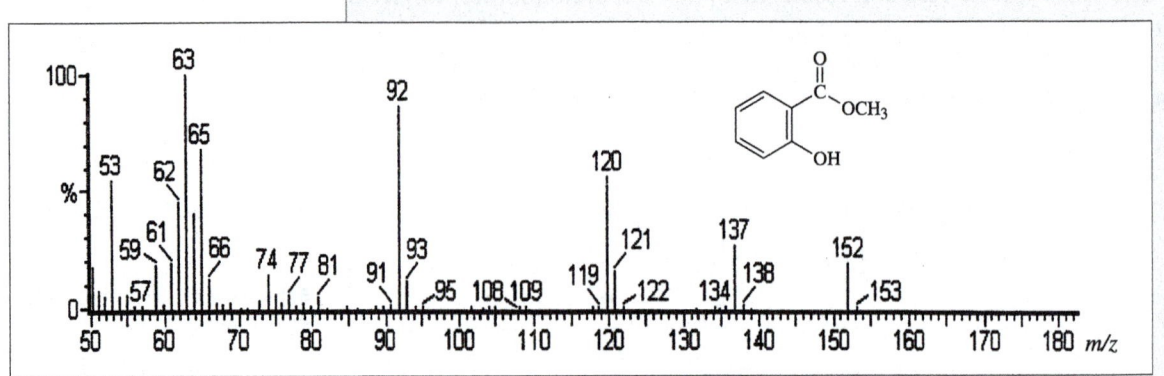

Figure 5.15 The EI MS of methyl 2-hydroxybenzoate (oil of wintergreen)

β-Cleavage next to the ester carbonyl is promoted by the *ortho* hydroxyl group, through a six-membered ring (Scheme 5.19). The radical cation produced, with m/z 120, is resonance stabilized and is observed as a major fragment in the EI MS.

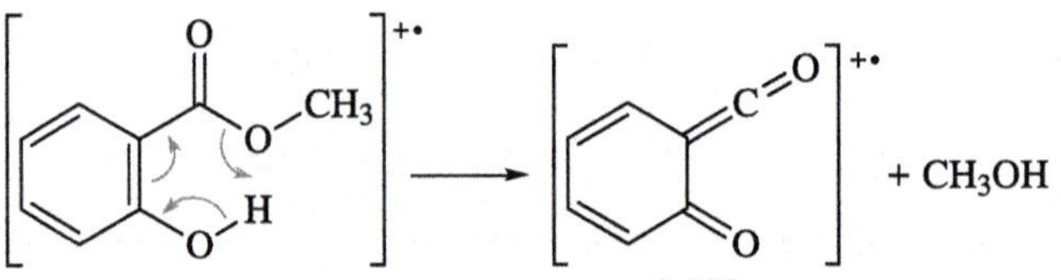

Scheme 5.19 β-Cleavage of methyl 2-hydroxybenzoate

5.4   Mass Analysis

Having generated the molecular ion (and, of course, any daughter ions which have arisen due to its fragmentation, as discussed in Section 5.3), the next step is the analysis of all the ions present. In order to do this, the ions are generally ejected from the ion source, by repulsion or attraction, into a mass analyser. As with the ionization process, there are a number of means of analysing the mass of ions, and we will again look only at those which are most commonly employed. We will not consider the detection of ions, since this is mainly electronics, and will concentrate purely on how the ion output from the ion source is analysed to give ultimately the mass spectrum in the form we are familiar with.

Since most mass spectrometers can usually utilize a range of ionization methods, the ion mass analysis methods generally characterize the type of instrument. The most common are:

- Magnetic sector mass spectrometers (and double-focusing mass spectrometers).
- Quadrupole mass filters.
- Ion trap mass spectrometers.
- Time-of-flight (TOF) mass spectrometers.
- Ion cyclotron resonance–Fourier transform (ICR-FT) mass spectrometers.

5.4.1 Magnetic Sector Mass Spectrometers (and Double Focusing Mass Spectrometers)

The very first mass spectrometers were magnetic sector instruments and, as the name suggests, these employ a magnetic field to analyse the ions produced in the ion source. The key equation (equation 5.1) relates the mass-to-charge ratio of the ion (m/z) to the magnetic field strength, B, the radius, r, of the circular path followed by the ions in a magnetic field, and the voltage used to accelerate the ions out of the ionization source, V:

$$\frac{m}{z} = \frac{B^2 r^2}{2V} \qquad (5.1)$$

Since the radius, r, is fixed by the geometry of the magnet, this means that by varying the magnetic field (B) while keeping the accelerating voltage (V) constant (or the other way round), we can scan through the mass spectrum. Ions of different m/z ratio have the required trajectory, and so pass through the collector slit, when the magnetic field satisfies equation (5.1) (see Figure 5.16). This arrangement has traditionally been the most used method for ion analysis, and gives a good separation of ions which differ by 1 mass unit, *e.g.* in Figure 5.10 the peaks at m/z 106 and 105.

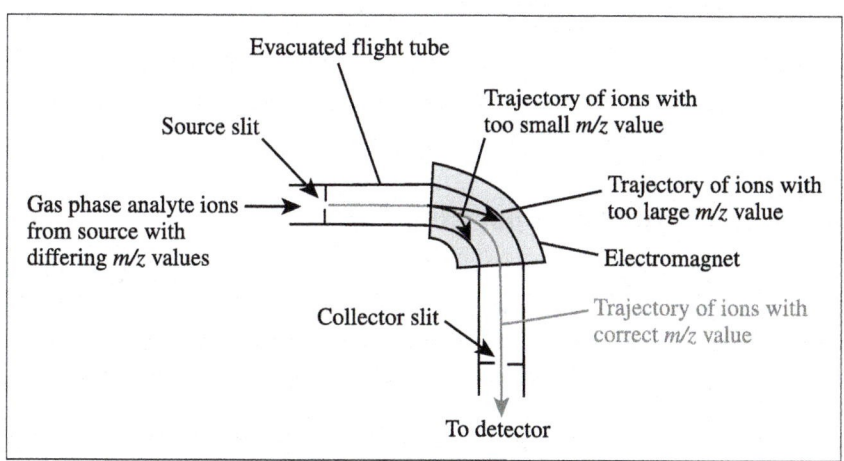

Figure 5.16 Schematic diagram of a magnetic sector mass spectrometer

Greater resolution can be achieved when a magnetic analyser is coupled with an electrostatic analyser (in a "double focusing" mass spectrometer). Using this combination of analysers, mass accuracies of around 1 part per million (ppm) can be obtained. However, magnetic sector instruments have relatively low sensitivity and are very expensive.

5.4.2 Quadrupole Mass Filters

A quadrupole analyser consists of two pairs of parallel rods. To one pair is applied a constant DC voltage (U) and an alternating (radiofrequency) voltage (V), and to the other pair is applied a DC voltage of opposite polarity and a radiofrequency voltage 180° out of phase with that on the other pair of rods. This arrangement acts as a mass filter, in a similar way to the magnetic analyser, but, in this case, separation of the ions requires the variation of U and V (whilst keeping the U/V ratio constant), thus changing the m/z ratio of the ions which achieve a stable trajectory in the field generated by the rods and so pass through the detector (Figure 5.17).

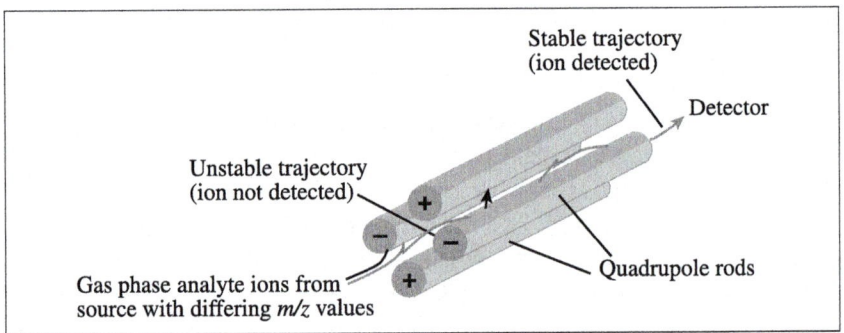

Figure 5.17 Schematic diagram of a quadrupole mass filter

Quadrupole mass filters can easily be combined with chromatographic techniques so are often the analysers used in GC-MS and LC-MS instruments (see Section 5.6). In general, quadrupole mass filters are often used to provide low-resolution spectra and are much cheaper and require less space than magnetic sector instruments.

5.4.3 Ion Trap Mass Spectrometers

The principles behind an ion trap mass spectrometer are similar to those of the quadrupole mass filter, except that the quadrupole field is generated within a three-dimensional cell using a ring electrode and no filtering of the ions occurs. All of the steps involved in the generation and analysis of the ions take place within the cell, and in order to detect the ions they must be destabilized from their orbits, by altering the electric fields, so that they exit the trap and are ejected in order of increasing m/z ratio to the detector.

Major benefits of ion traps are their compactness, the ease with which they are coupled to chromatographic techniques (they are frequently employed in LC-MS), their high sensitivity and the ease with which MS-MS (see Section 5.7) can be performed.

5.4.4 Time-of-flight Mass Spectrometers

In time-of-flight (TOF) spectrometers, as the name implies, the mass spectrum is generated by separating the ions according to the time it takes them to reach the detector. Unlike the other techniques we have met, this separation takes place in a region in which there is no applied magnetic or electric field ("the field-free region"). In a time-of-flight mass spectrometer, all ions of the same charge are given the same kinetic energy by accelerating them through a known potential difference.

The kinetic energy (KE) of the ions is given by equation (5.2). If the kinetic energy is constant, the ions with smaller masses will have greater velocities (and so take the shortest times to reach the detector), while those with greater masses will travel more slowly and so take longer to reach the detector. Ions with the same charge will therefore reach the detector in order of increasing mass.

$$KE = \tfrac{1}{2}mv^2 \qquad (5.2)$$

m = mass; v = velocity.

TOF mass spectrometers are among the most sensitive of mass analysers and can operate up to very high molecular masses (very low velocities).

5.4.5 Ion Cyclotron Resonance–Fourier Transform (ICR-FT) Mass Spectrometers

Once again, this technique utilizes an ion trap in which the ions are trapped within a cell which is situated within a strong magnetic field at right angles to the trapping plates. Ions in such a strong magnetic field undergo ion cyclotron resonance, and move in a circular orbit perpendicular to the magnetic field direction, at a frequency (the cyclotron frequency) which is dependent upon their m/z ratio. The ions can be excited by a pulse of radiofrequency voltage applied at their cyclotron frequency and, when this pulse is switched off, the movement of the ions generates an image current in the detector plates which decays with time due to collisions (free induction decay).

This process is somewhat similar to that involved in FT NMR (see Chapter 4) and we can acquire a number of scans, add them together, and perform a Fourier transform (which increases the signal-to-noise ratio). As we can measure frequencies very accurately, we can therefore measure the corresponding masses very accurately, and the ICR-FT MS is exceptionally useful in mass analysis ("high-resolution mass spectrometry"). It is also particularly suited to MS-MS (see Section 5.7).

5.5 Mass Spectral Data

5.5.1 Isotope Peaks

The **relative atomic mass** of an element is the *weighted mean* of the isotopic masses. The weighted mean is calculated from the masses of all the possible isotopes of the element, taking into account the natural relative abundance of each isotope.

When we calculate the relative mass of a molecule, *e.g.* in order to calculate the number of moles present in a weighed sample, we use the tables of average relative atomic masses, which take account of the percentage of each isotope in every sample and average out the molecular mass in the sample. However, mass spectral analysis gives the mass of each *individual* ion, with its particular combination of isotopes, rather than an average mass for all molecules present. Thus, the mass measured in a mass spectrometer will always differ from the average relative molecular mass of our compound (calculated using tables of average relative atomic masses) by an amount that is dependent on the mass of our compound, and this difference gets bigger as the mass of our compound increases. In order to calculate the relative mass of an analyte as measured in a mass spectrometer, we need to use a table of monoisotopic masses (Table 5.1).

Table 5.1 Relative masses and natural abundances for some commonly occurring elements

Isotope	Natural abundance (%)	Relative mass (to 4 d.p.)
^{1}H	100.00	1.0078
^{12}C	98.89	12.0000
^{13}C	1.11	13.0034
^{14}N	99.63	14.0031
^{15}N	0.37	15.0001
^{16}O	99.76	15.9949
^{17}O	0.04	16.9991
^{18}O	0.20	17.9992
^{19}F	100.00	18.9984
^{31}P	100.00	30.9738
^{32}S	95.02	31.9721
^{33}S	0.75	32.9715
^{34}S	4.21	33.9679
^{36}S	0.02	35.9671
^{35}Cl	75.73	34.9689
^{37}Cl	24.47	36.9659
^{79}Br	50.69	78.9183
^{81}Br	49.31	80.9163

The ability of the spectrometer to measure the mass of individual molecules gives rise to more than one peak for each molecule containing one or more atoms with isotopes. For example, when we examine the mass spectrum for a molecule with an m/z of 609, such as reserpine (Figure 5.18), we can see the smaller isotopic peaks at m/z 610 and 611, as well as the molecular ion peak (MH^+).

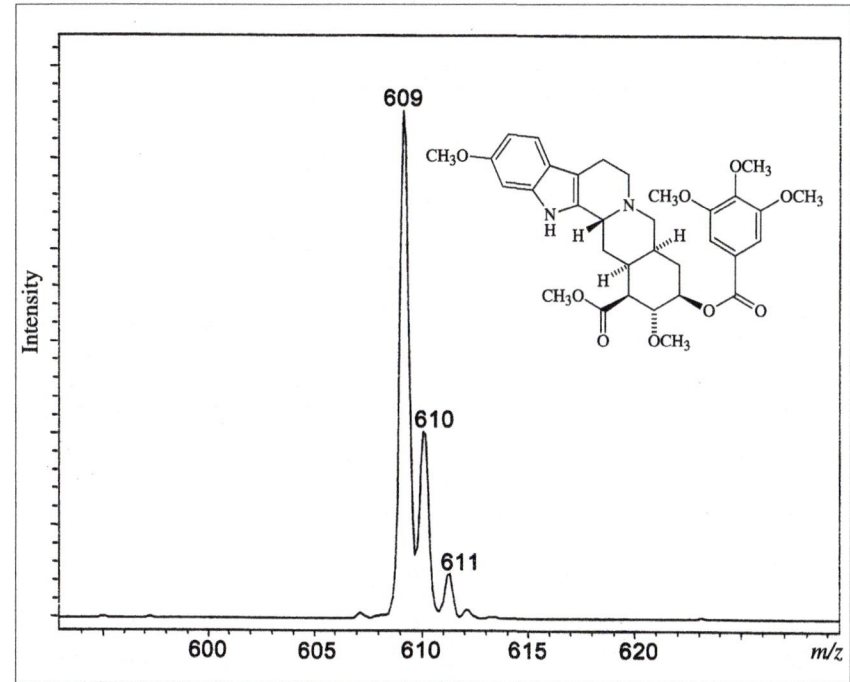

Figure 5.18 ESI mass spectrum of reserpine

The intensity of the isotopic peaks is related to the abundance of the isotope present and to the total number of atoms present in that molecule, *i.e.* it is related to the probability of finding that isotope in the molecule: the greater the abundance and the more atoms there are present, the greater the chance of finding an isotope and the more intense the isotopic peak.

For instance, carbon has two naturally occurring isotopes, ^{12}C with a natural abundance of 99.89% and ^{13}C with a natural abundance of 1.11%, so that roughly one in every 100 carbon atoms will be a ^{13}C. In a molecule containing 10 carbon atoms there are 10 chances of finding a ^{13}C atom, which adds up to a 1 in 10 chance that this particular molecule will contain one ^{13}C, so that for a C_{10} molecule we should see a peak corresponding to M + 1 with an intensity of 1/10 of that of the molecular ion. For small to medium sized molecules (*i.e.* <100 carbons) the most abundant peak is the one corresponding to exclusively ^{12}C atoms;

The peak at m/z 610 in Figure 5.18 represents those molecules of reserpine which incorporate 1 atom of ^{13}C somewhere in the structure. As there are 33 carbon atoms in reserpine, there is a $33 \times 1.11\%$ probability of there being a ^{13}C atom in any particular molecule of reserpine, which equals 36.6%; this explains the intensity of the m/z 610 peak. The peak at m/z 611 represents molecules of reserpine that incorporate two ^{13}C atoms: the peak height is again about 37% of the peak at m/z 610 (or 13.4% of the parent peak at m/z 609).

for molecules of very much higher masses (*e.g.* proteins) we find that the peak with the highest abundance contains a number of ^{13}C atoms. The same argument may be applied to all other elements present in any sample, and so what we see in any mass spectrometer is the isotope distribution present in the analyte.

The isotope patterns of chlorine and bromine are worth particular mention. Chlorine has two isotopes of mass 35 and 37, in a ratio of 75:25, respectively, while bromine has two isotopes of mass 79 and 81 in an approximately 50:50 ratio. If we examine the mass spectrum for 2-chlorobenzoic acid, with a molecular formula of $C_7H_5ClO_2$ (Figure 5.19), we can see peaks at $(MH+1)$ and $(MH+2)$ corresponding to the presence of the ^{13}C and ^{37}Cl isotopes, respectively.

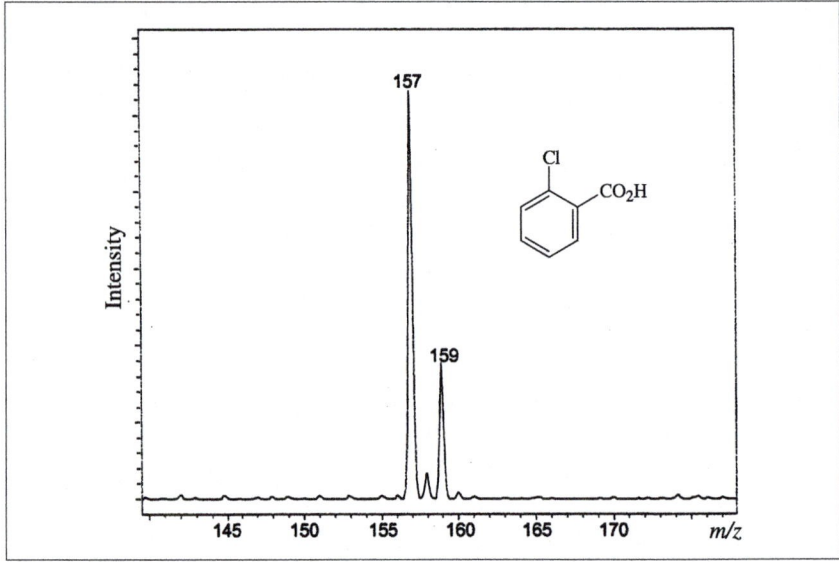

Figure 5.19 ESI mass spectrum of 2-chlorobenzoic acid

If you look carefully at Figure 5.19, you can see a peak at m/z 160. To what does this correspond?

The relatively intense isotope peaks separated by 2 mass units for Cl or Br provide a rapid indication of the presence of these elements in an analyte. Dichloro and dibromo compounds give similarly distinctive patterns, as shown in Figures 5.20 and 5.21. Other elements have similarly recognizable isotopic patterns, *e.g.* sulfur, although none is as distinctive as those for Cl and Br.

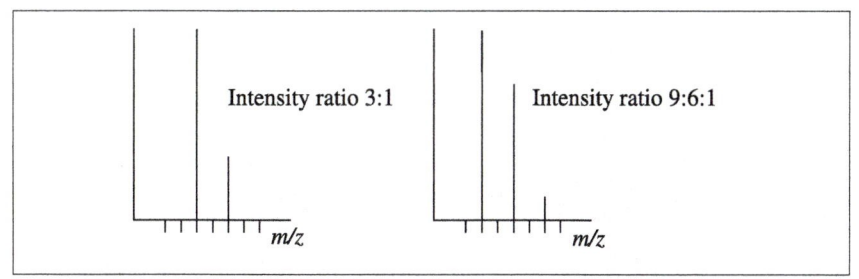

Figure 5.20 Isotope patterns for Cl and Cl_2

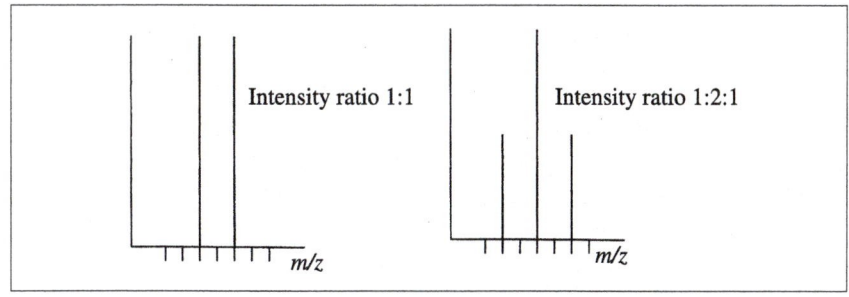

Figure 5.21 Isotope patterns for Br and Br_2

5.5.2 Mass Accuracy

Before we can discuss the issue of mass accuracy we must first consider what we mean by this term. If we examine Table 5.1 more closely, we see that the only element listed with an integral mass (*i.e.* a whole number) is ^{12}C, as defined by the IUPAC convention, and when we calculate the monoisotopic mass of an analyte we always find that it is not an integer. The decimal part of the mass is known as the mass defect because it causes the value to be non-integral. For example, when we use the values in Table 5.1 to calculate the relative mass for the molecular formula $C_{10}H_{10}N_3O_2$, we obtain a mass of 204.0771, where the figure 0.0771 is the "mass defect" of this compound. You can see that for isomers the mass defect will be exactly equal, and we require other techniques, such as MS-MS or fragmentation pattern analysis, to help us to distinguish these structures (see Section 5.7). We can, however, exploit the mass defect to our advantage if we can measure our mass (m/z) with sufficient accuracy. As there are only a certain number of isotopic combinations that can give rise to a particular value of the mass defect, we now have a potential means of determining molecular formulae. There is obviously a range of different molecular formulae that can fit any mass. For example, if we take the two formulae $C_{12}H_{16}N_2O_4$ and $C_{11}H_{12}N_2O_5$ we can see that they both have the same relative molecular mass of 252.

However, when we calculate the "accurate" masses of these compounds based on the masses given in Table 5.1, we see that they are

IUPAC is the acronym for the International Union of Pure and Applied Chemistry, the body that oversees and agrees international standards in chemistry, *e.g.* units or the nomenclature of organic compounds.

The term used to describe two compounds with the same overall relative molecular mass but with different molecular formulae is **Isobaric species**.

252.1106 and 252.0743, respectively. Thus, the mass defects for these two species are different. If we could measure these masses to an accuracy of four decimal places, we could compare the measured mass against the theoretical mass and thereby determine the elemental composition of our sample. This process is known as mass analysis (but is often, wrongly, referred to as high-resolution mass spectrometry).

The scale we use to measure mass accuracy is parts per million (ppm), which is a relative unit obtained by dividing the absolute error in the mass measurement by the relative molecular mass according to equation (5.3):

$$\text{Mass error in ppm} = \frac{(\text{absolute mass error}) \times 10^6}{\text{relative molecular mass}} \tag{5.3}$$

Thus, if we measure the mass of a compound with a relative molecular mass of 1000 to an accuracy of 0.001 mass units, we have measured its mass with an accuracy of 1 ppm.

A very simple and useful rule to aid in the interpretation of the MS data of organic compounds relates to the fact that nitrogen has an even mass but an odd valency, whereas the valency and mass for the most abundant isotope of other elements are either both odd or both even. This gives us a very simple rule that we can immediately apply to any mass spectral data: the nitrogen rule, which is given in Box 5.2.

> *You may like to consider how "the nitrogen rule" arises for neutral organic amines.*

Box 5.2 The Nitrogen Rule

A neutral organic compound, composed of the most abundant isotopes of the elements present, with an even number of nitrogen atoms must have an even relative molecular mass, but a compound with an odd number of nitrogen atoms must have an odd relative molecular mass.

5.6 Hyphenated Mass Spectrometry Methods

A great number of chromatographic techniques are available to the analyst for separating the components present in a mixture. Of these, only two are commonly interfaced to MS systems: high-performance liquid chromatography (HPLC or LC) and gas chromatography (GC).

In theory, the MS system used in these techniques is no different from any other detection system (*e.g.* a UV detector or flame ionization detection), except that it is a good deal more expensive. However, its main advantage over other detectors is the additional information about the analyte that it provides and the greater sensitivity it offers.

The power of GC-MS and LC-MS techniques lies in their *potential* ability to take a mixture of many different compounds and provide structural information on each component. There is some overlap in the types of compounds that may be analysed using these two techniques, but, in the main, the role of both techniques may be considered to be complementary. In which circumstances then do we use GC-MS and in which do we use LC-MS? As a general rule, if the analyte is non-polar and therefore volatile, we would probably use GC-MS. On the other hand, if the analyte is polar and/or non-volatile, the most appropriate technique will probably be LC-MS.

5.6.1 GC-MS

GC, as a chromatographic technique, is naturally suited to MS because it produces analyte molecules that are already in the gas phase (a requirement for any MS analysis). Interfacing a GC system to an MS instrument is, therefore, relatively straightforward as the compounds eluting from a GC column have already been volatilized and merely require separation from the carrier gas and ionization (usually by EI or CI; Section 5.2.1) before mass analysis.

The process of recording a mass spectrum is repeated many times during a GC run (*e.g.* at a rate of 1 scan per second), and by recording the number of ions reaching the detector we obtain a total ion chromatogram (TIC). A TIC is very similar in appearance to a normal chromatographic trace produced when using any other sort of GC detector, except that we have a mass spectrum available for each point on the chromatogram.

EI, as a relatively harsh ionization technique, will always cause fragmentation of the molecular ion and hence generate a more or less unique fragmentation fingerprint of the analyte. It is thus possible to perform a search against a library of standard MS spectra and, under the right circumstances, identify the analyte. This is a process that must be undertaken with care as false matches are often produced, but the ability of this technique to identify an unknown analyte from a known compound library, as a single shot experiment, is virtually unparalleled. The standard library of EI mass spectra is that produced by the American National Institute of Standards and Technology (NIST), which currently contains mass spectra for >120,000 compounds.

This technique has found great application in forensic science, *e.g.* in the detection of drugs in athletes or racehorses.

5.6.2 LC-MS

LC-MS, as a technique, is very much dependent upon ionization (and ion vaporization) techniques that are suited to LC conditions, *i.e.* techniques where a relatively large solvent flow can be accommodated, which restricts us to just two ionization methods: electrospray ionization (ESI) and atmospheric pressure chemical ionization (APCI). Both techniques are very similar in their modes of operation (see Section 5.2.1), relying on the formation of a spray from a solvent flow at atmospheric pressure, and hence they are ideally suited to use in LC-MS applications.

The power of LC-MS over MS as a single technique is the ability to separate out the components of a mixture and then provide MS data on each. The use of accurate mass LC-MS is increasing as it enables the molecular formula of the analyte to be determined as an additional piece of information in the structural elucidation jigsaw.

5.7   MS-MS

There are circumstances where it is advantageous to know more information about our analyte than just its relative molecular mass, *e.g.* in cases where we have mixtures of isomers in which LC-MS will be unable to provide any further useful information. One way to obtain this useful structural information is to fragment the molecular ion, and this may be carried out either in the MS source or within the mass analysis device. For various technical reasons the second of these is a superior fragmentation technique and will always be used, when available, in preference to the in-source version. The term used to describe fragmentation in the mass analyser is MS-MS. Fragments formed using MS-MS techniques can be different from those formed using EI or CI, but they nonetheless follow a similar set of relatively simple rules (*e.g.* loss of H_2O from aliphatic alcohols) and a particular fragmentation pattern is often characteristic of a particular structure. For MS systems based on a trapping principle, the MS-MS process does not need to be halted at the first stage, and further fragmentation of the fragments themselves maybe carried out — a process described by the term MS^n.

Summary of Key Points

1. Mass spectrometry is used to measure the relative mass of molecules and requires the generation of analyte ions, followed by mass analysis and ion detection.

2. Ion generation can be achieved in a number of ways: electron impact (EI) ionization, chemical ionization (CI), fast atom bombardment (FAB), matrix assisted laser desorption ionization (MALDI), electrospray ionization (ESI) and atmospheric pressure chemical ionization (APCI) are the most common methods.

3. There are many common fragmentation processes which can help to explain and predict ion fragments found in EI spectra. In general, these processes are promoted by the stability of the carbocation fragments produced and by six-membered transition states in the rearrangements of ions.

4. Mass analysis is the process by which the relative mass of the ions is measured, and characterizes the type of instrument.

5. Mass spectrometers measure the relative mass of an individual ion, with its particular combination of isotopes. We therefore use the relative atomic masses of the most abundant isotopes to calculate the accurate relative molecular mass of an ion.

6. The isotope patterns of chlorine and bromine are particularly distinctive, such that analytes containing one or two chlorine atoms or one or two bromine atoms can be readily distinguished.

7. Mass analysis is the process by which the "mass defect" can be used to determine the molecular formula of an ion.

Problems

5.1. If no other ionization method is available, how could the extent of fragmentation be decreased in EI?

5.2. In CI, which reagent gas would you expect to give rise to a smaller degree of fragmentation, and why?

5.3. The Friedel–Crafts alkylation of benzene with 1-chloropropane in the presence of aluminium trichloride gives, after chromatography, the isomeric products **A** (76%) and **B** (24%). Predict the products of the fragmentation of both **A** and **B** and, by comparing them with the mass spectra given in Figures 5.22 and 5.23, deduce which is the major, **A**, and the minor, **B**, product of this reaction.

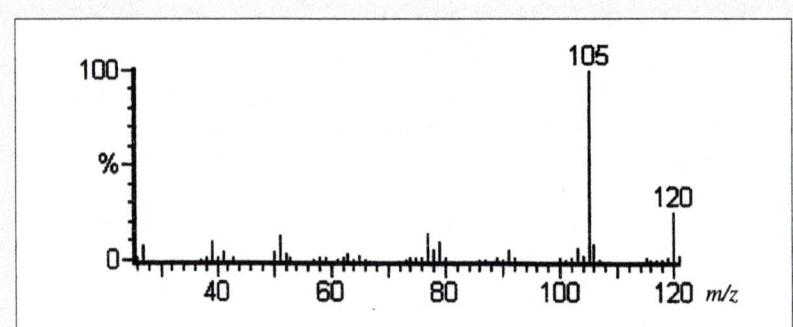

Figure 5.22 EI MS of **A**

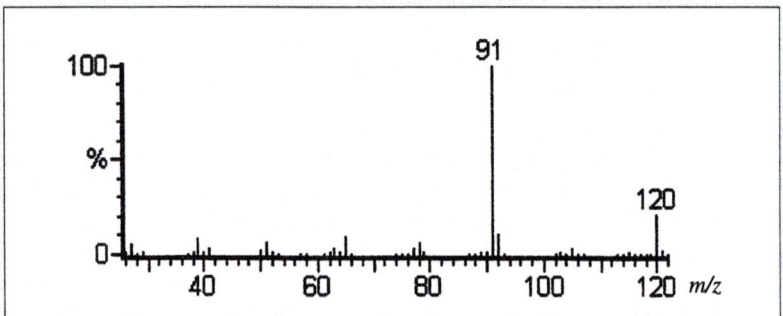

Figure 5.23 EI MS of **B**

5.4. The molecular formula of reserpine is $C_{33}H_{40}N_2O_9$ and that of the MH^+ ion, formed in electrospray MS, is $C_{33}H_{41}N_2O_9$. Using the isotopic masses given in Table 5.1, calculate (a) the exact mass of the MH^+ ion and (b) the exact mass and relative intensity (compared to the MH^+ ion) of the $[MH+1]^+$ peak, which is due to the ^{13}C isotopic abundance (*i.e.* $^{12}C_{32}^{13}C_1H_{41}N_2O_9$).

5.5. The mass spectra of the homologous series of alkyl 2-hydroxybenzoates (Figure 5.24) contain common daughter ion peaks at m/z 121, 120 and 93, whilst the ethyl $[R=CH_2CH_3]$, *n*-propyl $[R=CH_2CH_2CH_3]$ and isopropyl $[R=CH(CH_3)_2]$ esters also contain a daughter ion peak at m/z 138. Suggest structures for all of these daughter ions and give mechanisms for their formation.

Figure 5.24 The homologous series of alkyl 2-hydroxybenzoates

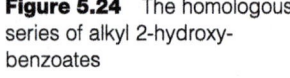

$R = CH_3, CH_2CH_3, CH_2CH_2CH_3, CH(CH_3)_2$

5.6. The molecular ion in the electrospray mass spectrum of the Pfizer anti-depressant Lustral (sertraline hydrochloride), $C_{17}H_{18}NCl_3$, contains the peaks shown in Figure 5.25. Explain the relative intensities of these peaks (given in brackets above each peak).

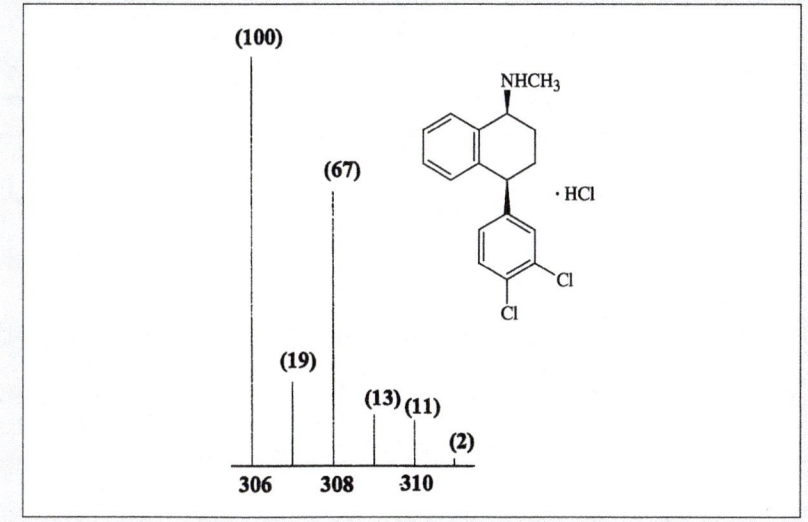

Figure 5.25 The ESI MS of sertraline hydrochloride

6

Structure Elucidation Using All of the Spectroscopic Information Available

Aims

The aim of this chapter is to illustrate, using a Worked Problem, a systematic method for the identification of an unknown organic substance from its spectra (UV-Vis, IR, ^{1}H, ^{13}C and 2-D NMR and mass), and then to give you some practice with examples designed to highlight the advantages of the different types of spectra available.

Worked Problem 6.1

Q Using the information obtained from the spectra shown in Figures 6.1–6.7, and the molecular formula of $C_{13}H_{18}O_2$, deduce the structure of **A**.

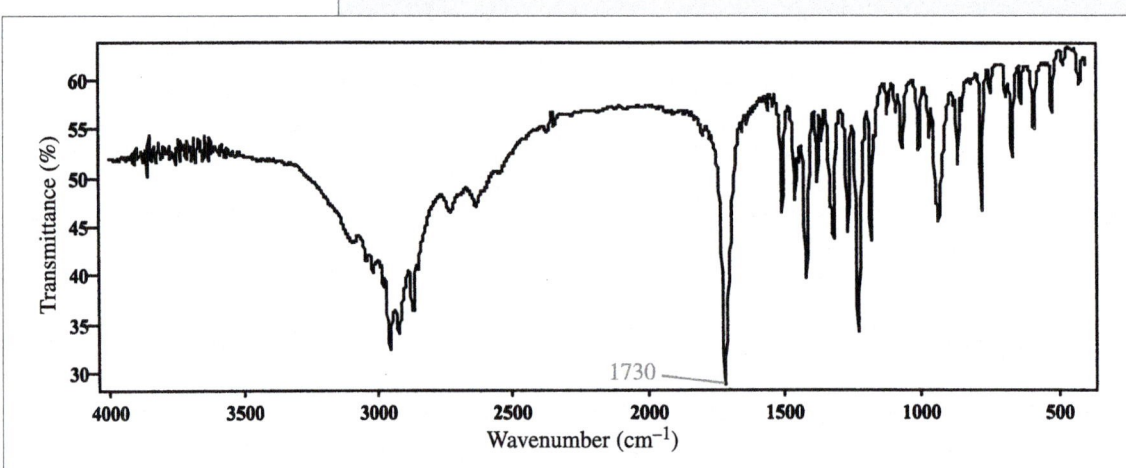

Figure 6.1 IR spectrum of unknown **A** (KBr disc)

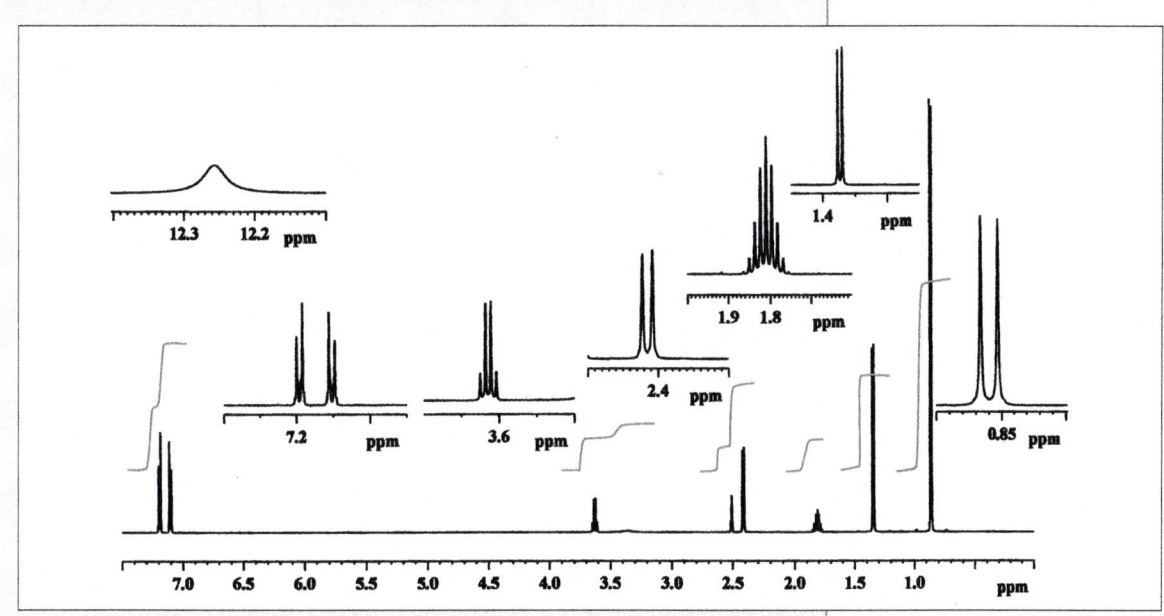

Figure 6.2 500 MHz ^{1}H NMR spectrum of unknown **A** in DMSO-d_6 (δ_H 2.51; δ_H H$_2$O 3.37): δ_H 0.86 (6H, d, J = 6.6 Hz), 1.34 (3H, d, J = 7.1 Hz), 1.81 (1H, 9 lines, J = 6.6 Hz), 2.41 (2H, d, J = 6.6 Hz), 3.63 (1H, q, J = 7.1 Hz), 7.10 (2H, d, J = 8.1 Hz), 7.19 (2H, d, J = 8.1 Hz), 12.25 (1H, broad s)

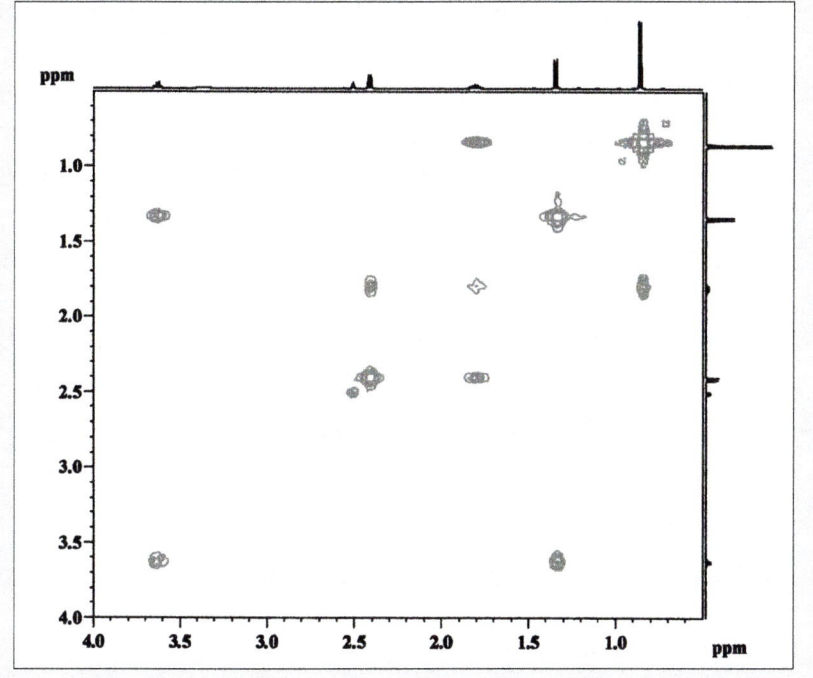

Figure 6.3 High-field region of the 500 MHz HH COSY spectrum of unknown **A** in DMSO-d_6

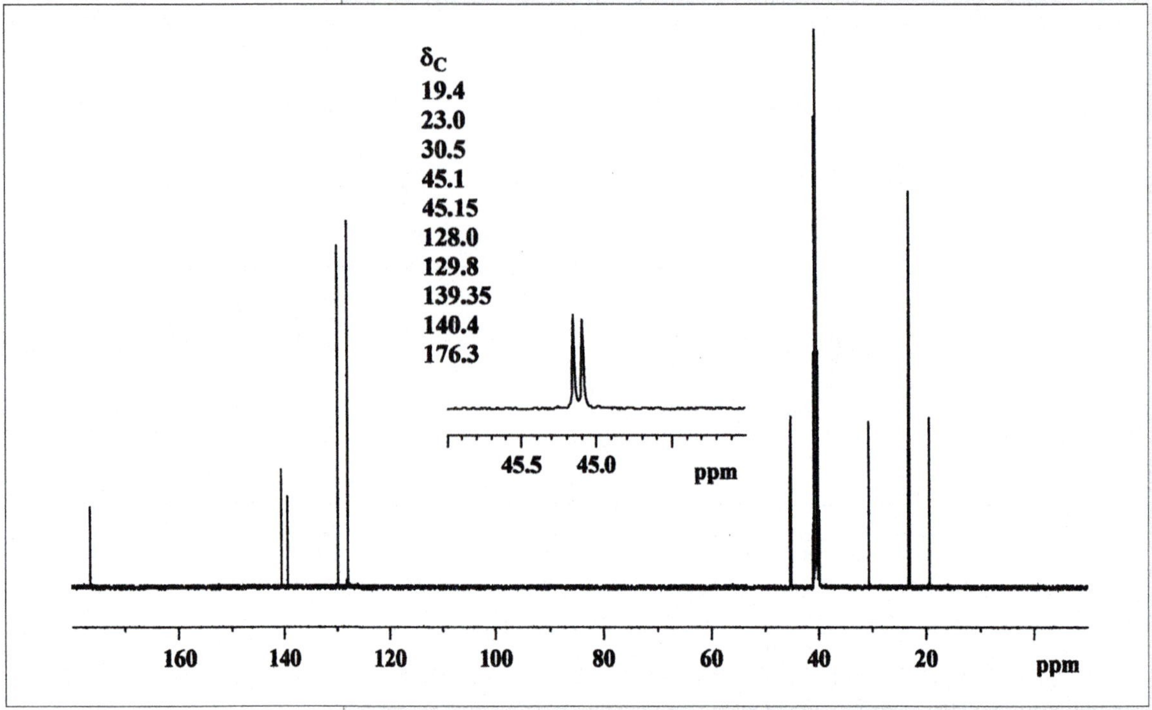

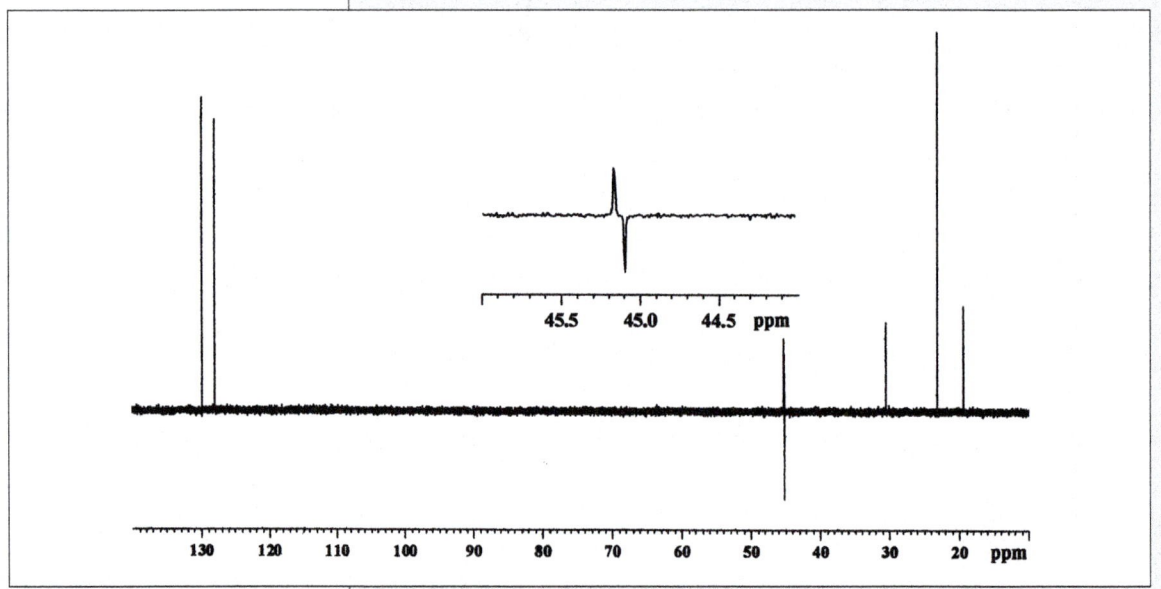

δ_C
19.4
23.0
30.5
45.1
45.15
128.0
129.8
139.35
140.4
176.3

Figure 6.4   125 MHz proton-decoupled ^{13}C NMR spectrum of unknown **A** in DMSO-d_6 (δ_C 40.5)

Figure 6.5   125 MHz DEPT 135 ^{13}C NMR spectrum of unknown **A** in DMSO-d_6

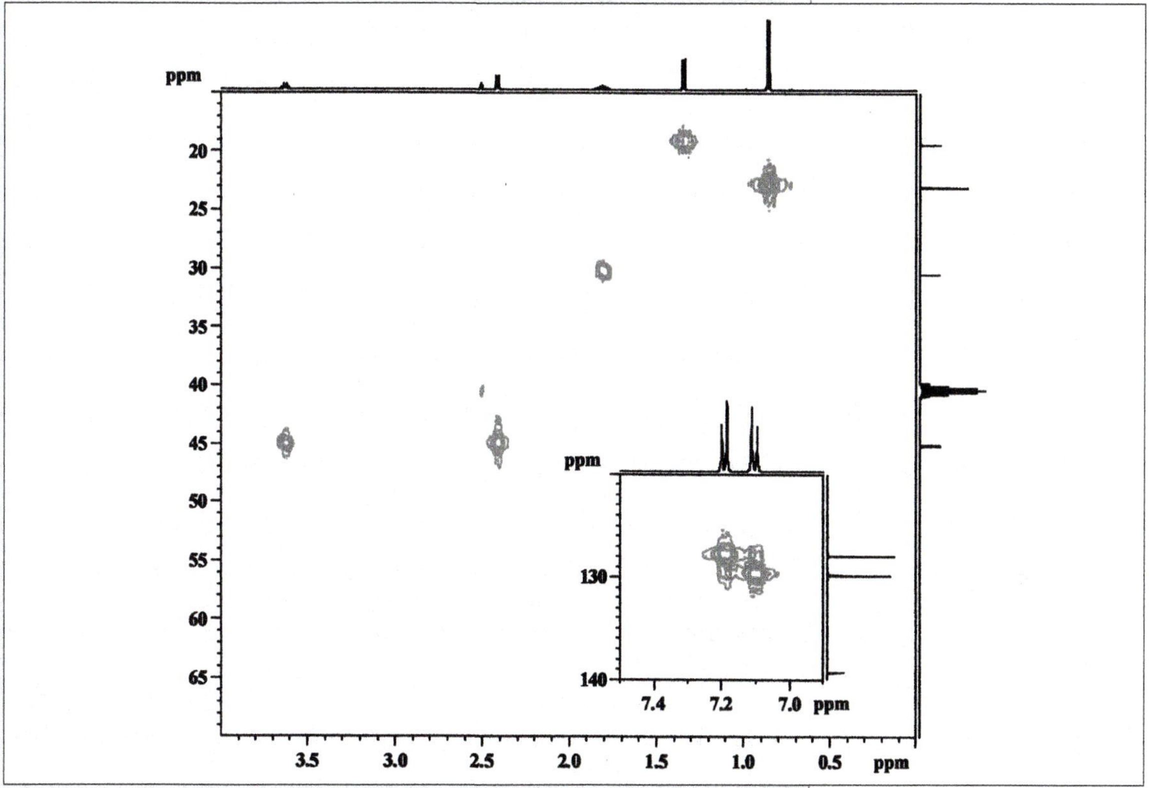

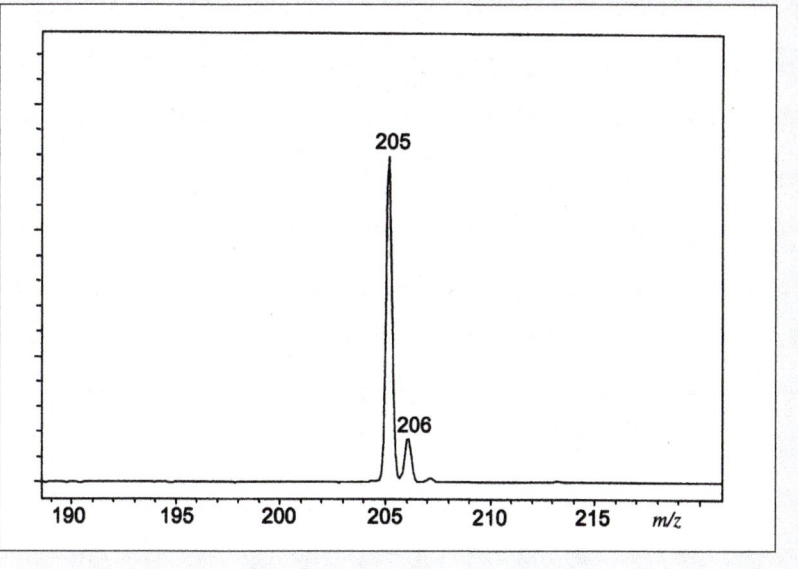

Figure 6.6 HMQC [$^1J(^{13}C—^1H)$] spectrum of unknown **A** in DMSO-d_6 (the aromatic region, shown in the bottom right to save space, does not hide any peaks)

Figure 6.7 Negative ion ESI mass spectrum of unknown **A**

$$\Omega = (C + 1) - \left[\frac{(H^* - N)}{2}\right]$$

where C = number of carbon atoms, H* = number of hydrogens or halogens, and N = number of nitrogens.

A Knowing the molecular formula of **A**, we recommend that the first step should be to calculate the number of double bond equivalents in the molecule, using equation (1.3), and then to use this as an indication of the number of functional groups present:

$$\Omega = DBE = (13 + 1) - (18/2) = 5$$

If we do this, we find that **A** contains five double bonds or their equivalents, and we should now try to identify these using the IR spectrum (Figure 6.1). The most notable feature of this spectrum is the strong carbonyl (C=O) absorption at 1730 cm^{-1}, which is in the region corresponding to the absorptions for acids, esters, aldehydes and ketones, but there is also a suggestion of a broad band (2900–2600 cm^{-1}) for an O–H group, which, when taken together with the carbonyl absorption, would suggest a carboxylic acid (–CO$_2$H). Below 1000 cm^{-1} there are a number of unsaturated out-of-plane bending absorptions, suggesting the presence of an aromatic ring, and this can be confirmed by a quick look at the ^{1}H NMR spectrum (Figure 6.2), in which there are two signals (at δ 7.10 and 7.19), each corresponding to 2H, with the splitting pattern characteristic of a 1,4-disubstituted (*para*) benzene ring (–C$_6$H$_4$–). The five double bond equivalents have therefore been identified as a C=O group and a benzene ring (4 DBE).

A fuller analysis of the ^{1}H spectrum should now help us to identify the structural fragments remaining, which, if we have been correct in our analysis so far, should correspond to C$_6$H$_{13}$ and contain no DBEs. The spectral data from Figure 6.2 are analysed in Table 6.1, along with the conclusions that can be drawn from the splitting patterns and the integrals for the peaks. Using the HH COSY spectrum (Figure 6.3) we can see that all of the expected cross-peaks are present, confirming that **A** contains the structural fragments shown in Figure 6.8, and it is not a giant leap from these to come up with the proposed structure for **A** as Ibuprofen, based upon these fragments. Notice that, as expected, the vicinal (^3J) coupling constant in the aromatic ring is much greater than those in the aliphatic parts of the molecule, and this is mostly due to the shorter carbon–carbon bonds in the benzene ring.

Table 6.1 Analysis of the ^{1}H NMR and HH COSY NMR spectra of unknown **A**

Chemical shift, δ	Integral	Splitting pattern	Structural group
0.86	6H	d, $J = 6.6$ Hz	$(CH_3)_2CH-$
1.34	3H	d, $J = 7.1$ Hz	CH_3CH-
1.81	1H	9 lines, $J = 6.6$ Hz	$(CH_3)_2CHCH_2-$
2.41	2H	d, $J = 6.6$ Hz	$-CH_2CH-$
3.63	1H	q, $J = 7.1$ Hz	$CH_3\mathbf{CH}-$
7.10	2H	d, $J = 8.1$ Hz	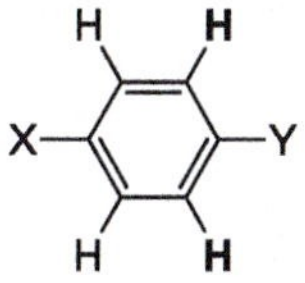 X—⟨benzene ring⟩—Y
7.19	2H	d, $J = 8.1$ Hz	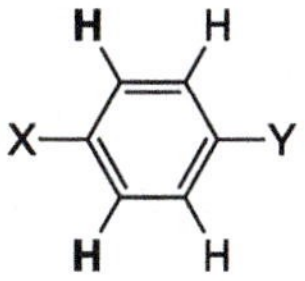 X—⟨benzene ring⟩—Y
12.25	1H	broad s	$-CO_2H$

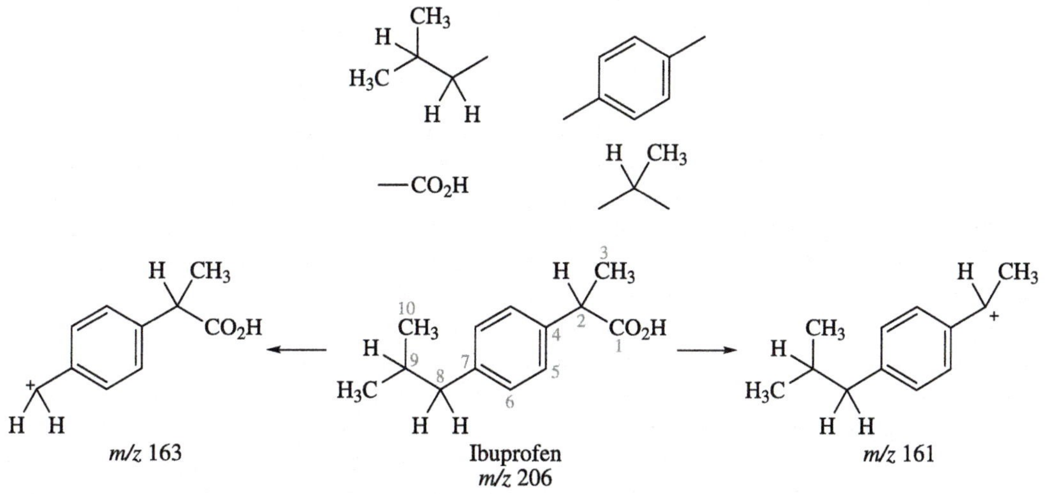

Figure 6.8 Fragmentation pattern for unknown **A**

(structures shown: m/z 163, Ibuprofen m/z 206, m/z 161)

It only remains for us to confirm this structure using the proton-decoupled ^{13}C (Figure 6.4), DEPT 135 (Figure 6.5) and HMQC (Figure 6.6) data; an analysis of these spectra is given in Table 6.2. In order to fully assign the aromatic carbon atoms (C4–C7), we could employ an HMBC spectrum, but this is not essential for the confirmation of this structure.

Table 6.2 Analysis of the proton-decoupled ^{13}C, DEPT 135 and HMQC spectra of unknown **A**

Chemical shift, δ	Assignment	C–H correlation, δ	Carbon
19.4	CH_3	1.34	3
23.0	CH_3	0.86	10
30.5	CH	1.81	9
45.1	CH_2	2.41	8
45.15	CH	3.63	2
128.0	CH	7.19	5 or 6
129.8	CH	7.10	5 or 6
139.35	quat.	–	4 or 7
140.4	quat.	–	4 or 7
176.3	quat.	–	1

The ESI mass spectrum confirms the $[M–H]^-$ ion as m/z 205 (Figure 6.8).

As you might have expected beforehand, the NMR spectra have been the most useful in determining this structure, but they should not be used in isolation, since a combination of all the techniques provides a very powerful tool for structural elucidation.

Problems

6.1. Unknown **B**, an extract of sassafras which is used in the manufacture of perfumes, has a molecular formula of $C_8H_6O_3$. Using the data in Table 5.1 on p. 142, calculate the expected accurate mass of **B** (MH^+). Using the approach outlined above, use the information in Figures 6.9–6.15 to deduce the structure of **B**.

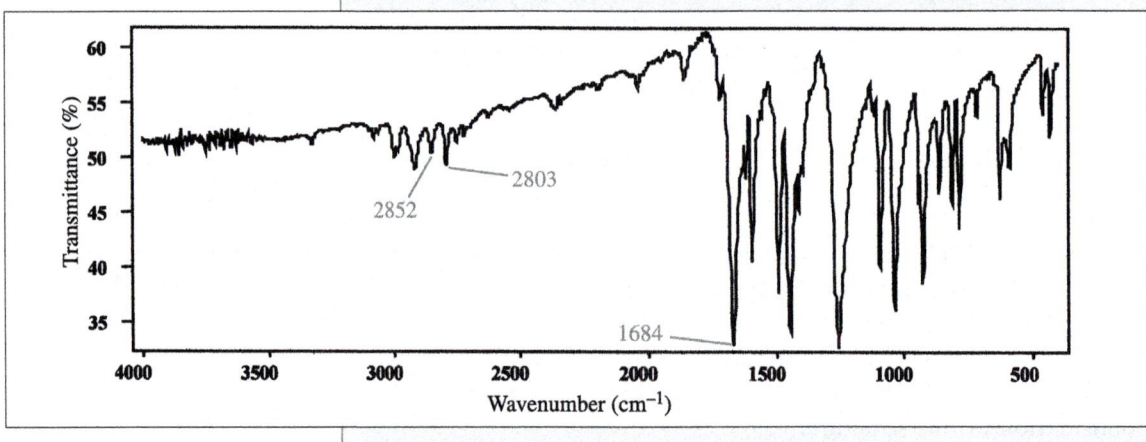

Figure 6.9 IR spectrum of unknown **B** (KBr disc)

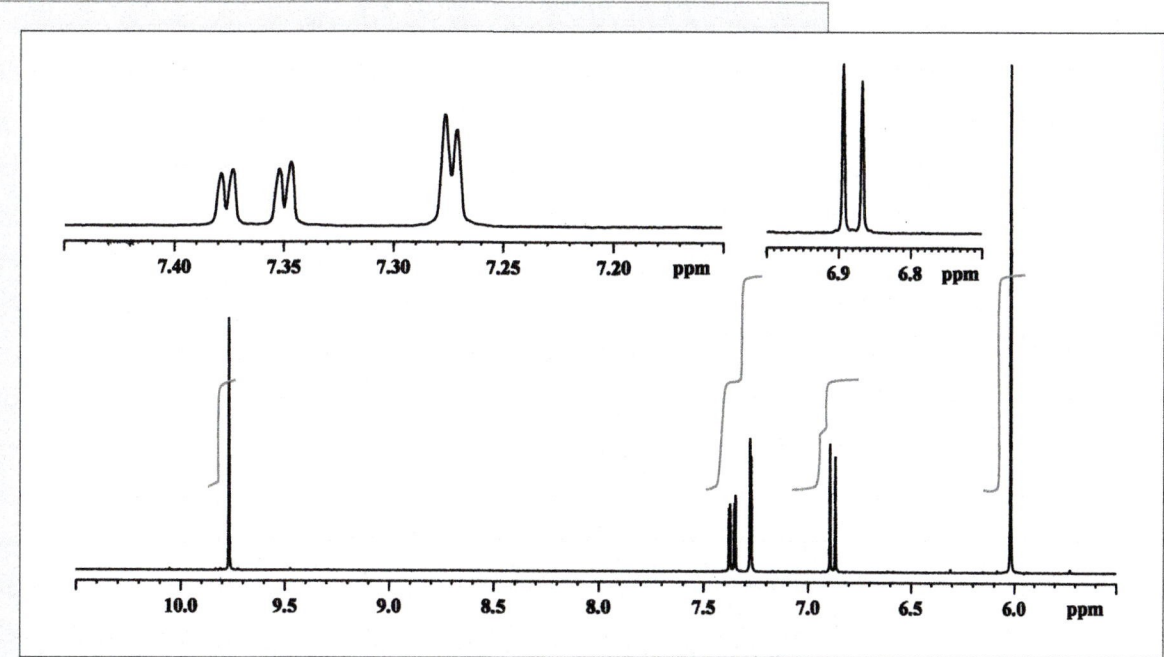

Figure 6.10   300 MHz ^{1}H NMR spectrum of unknown **B** in CDCl$_3$: δ_H 6.04 (2H, s), 6.89 (1H, d, $J = 7.95$ Hz), 7.28 (1H, d, $J = 1.6$ Hz), 7.37 (1H, dd, $J = 7.95, 1.6$ Hz), 9.77 (1H, s).

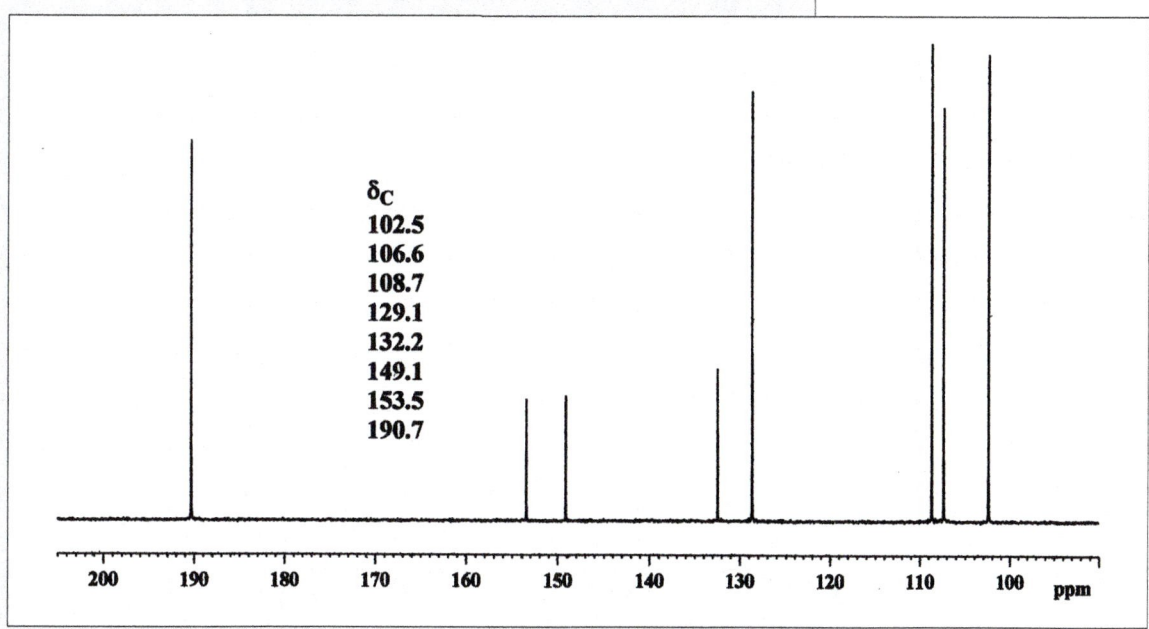

δ_C
102.5
106.6
108.7
129.1
132.2
149.1
153.5
190.7

Figure 6.11   75 MHz proton-decoupled ^{13}C NMR spectrum of unknown **B** in CDCl$_3$

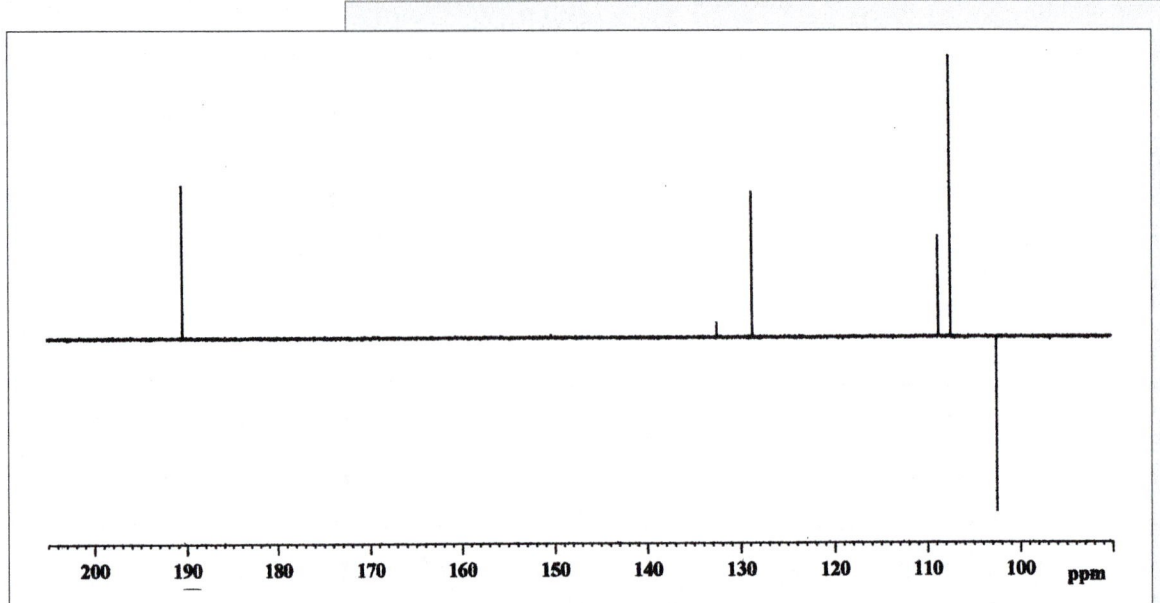

Figure 6.12   75 MHz DEPT 135 ^{13}C NMR spectrum of unknown **B** in CDCl$_3$

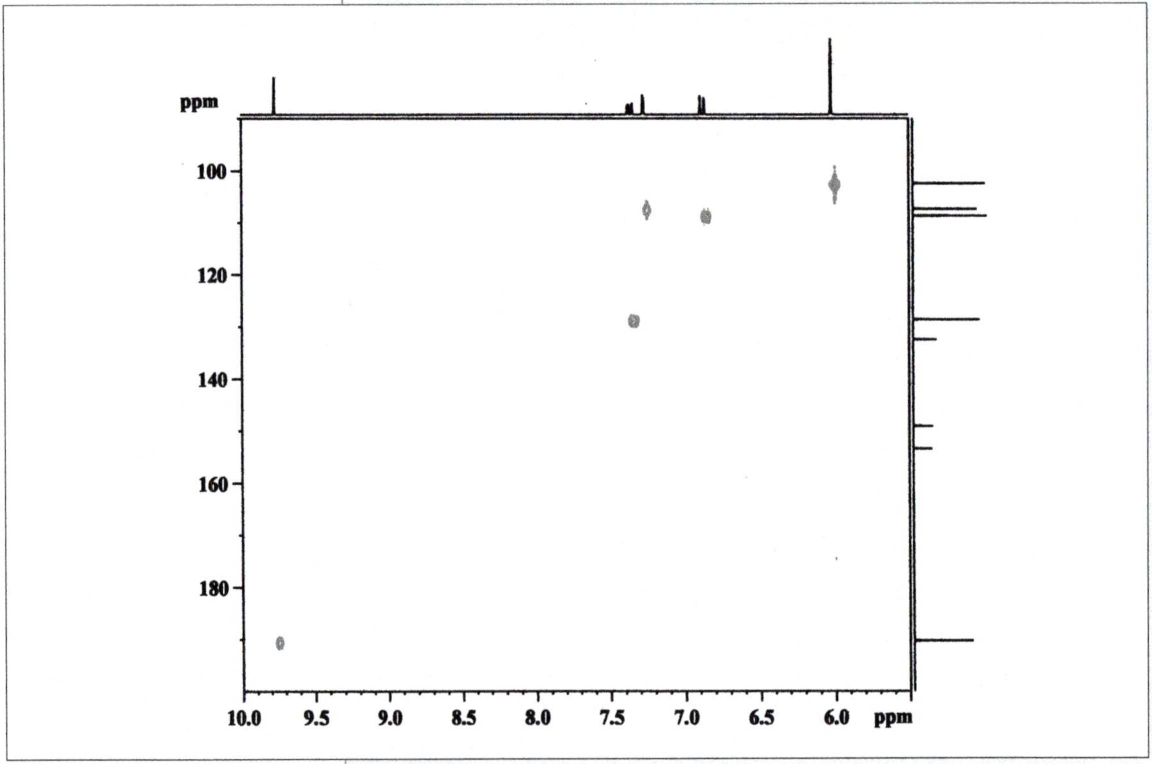

Figure 6.13   HMQC [$^1J(^1\text{H}-^{13}\text{C})$] NMR spectrum of unknown **B** in CDCl$_3$

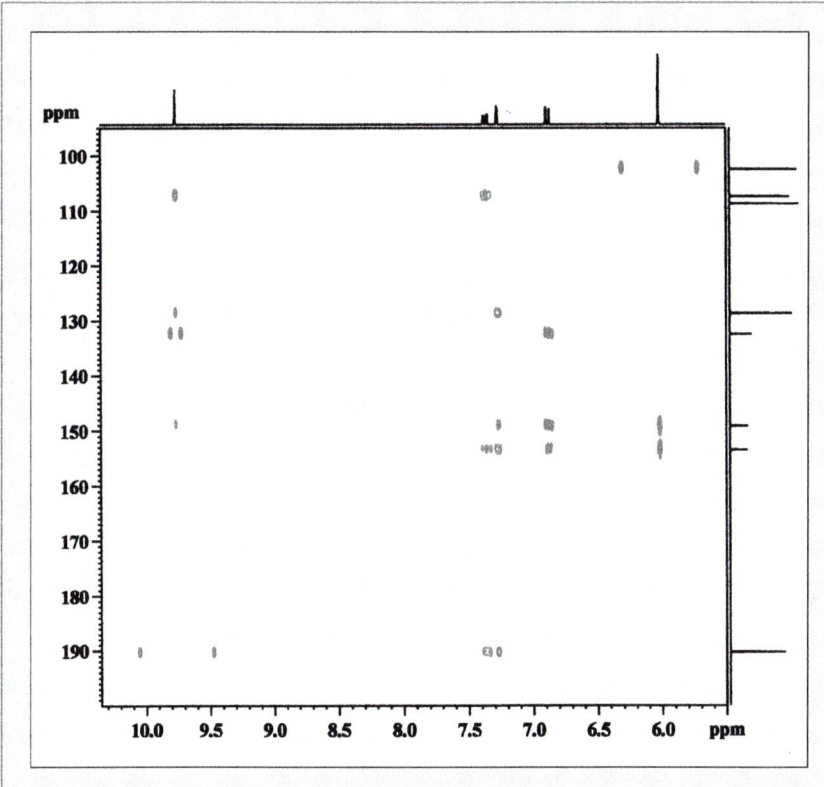

Figure 6.14 HMBC [$^2J(^1H-^{13}C)$ and $^3J(^1H-^{13}C)$] NMR spectrum of unknown **B** in CDCl$_3$

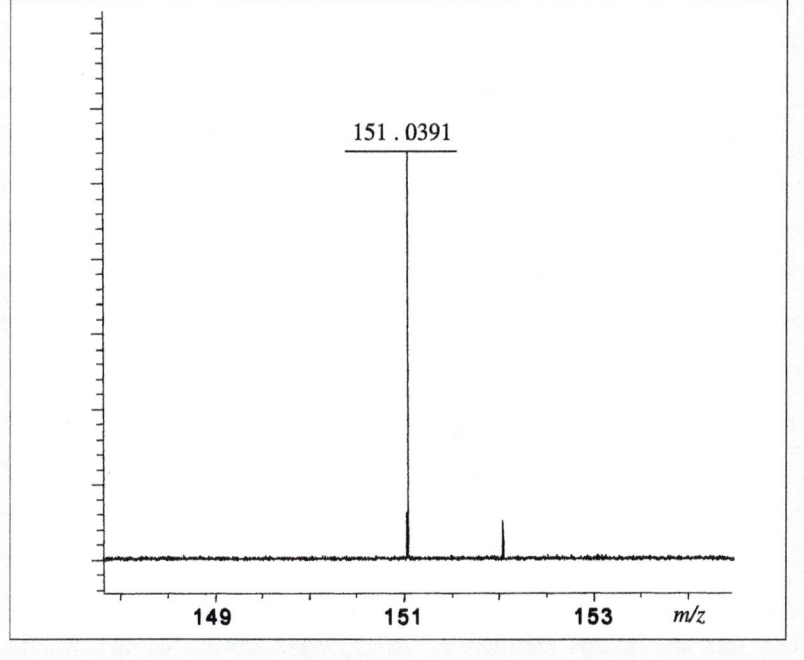

Figure 6.15 Accurate mass APCI+ mass spectrum of unknown **B**

6.2. Unknown **C**, which is also used in the manufacture of perfumes, has a molecular formula of $C_8H_{14}O$. Using the data in Table 5.1 on p. 142, calculate the expected accurate mass of **C** (MH^+). Using the data in Figures 6.16–6.23, deduce the structure of unknown **C**.

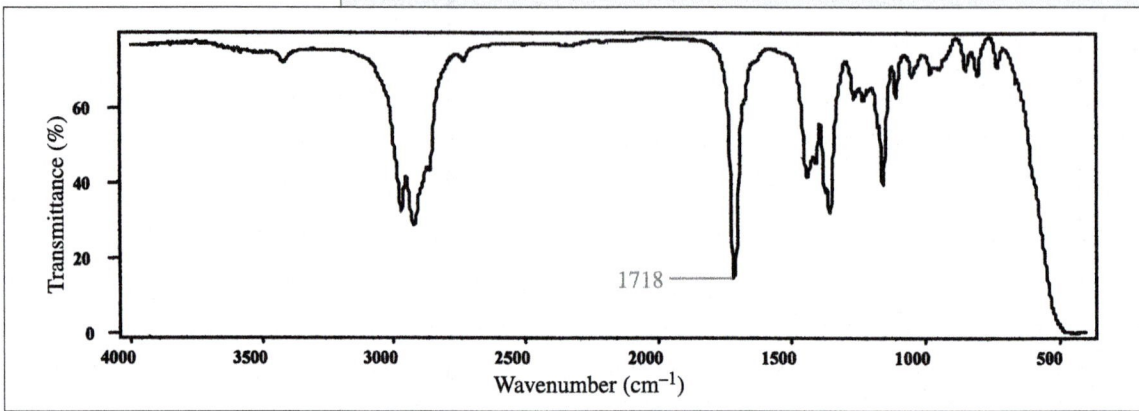

Figure 6.16 IR spectrum of unknown **C** (liquid film)

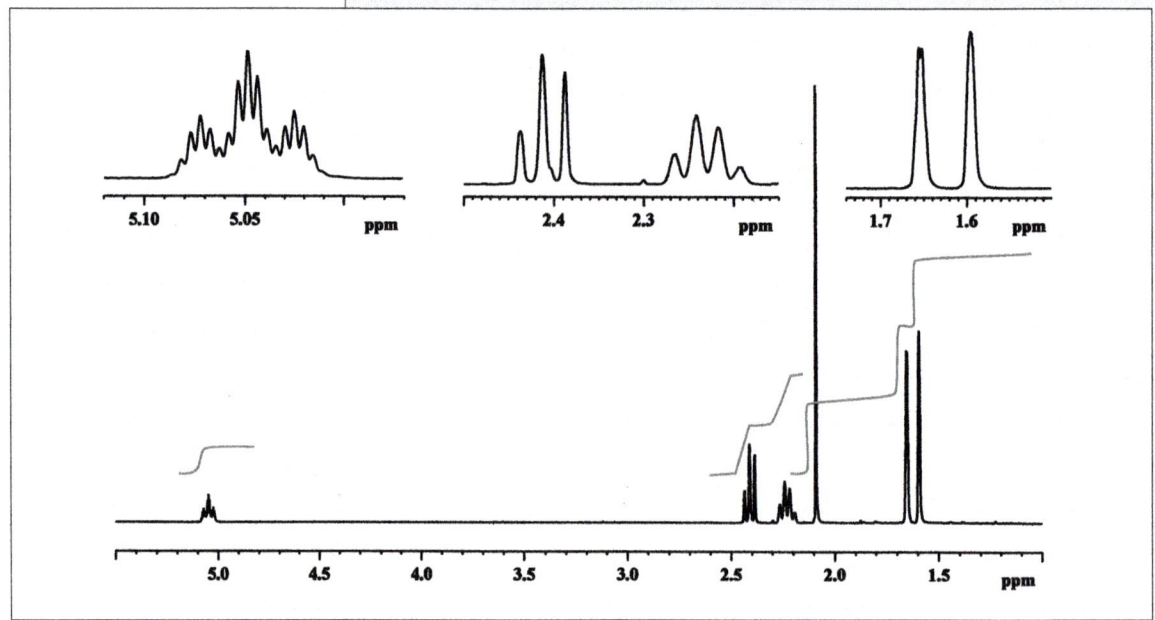

Figure 6.17 300 MHz ^{1}H NMR spectrum of unknown **C** in $CDCl_3$: δ_H 1.59 (3H, d, $J = 1.4$ Hz), 1.67 (3H, d, $J = 1.4$ Hz), 2.09 (3H, s), 2.24 (2H, q, $J = 7.2$ Hz), 2.41 (2H, t, $J = 7.2$ Hz), 5.05 (1H, 3 × 7 lines, $J = 7.2$, 1.4 Hz)

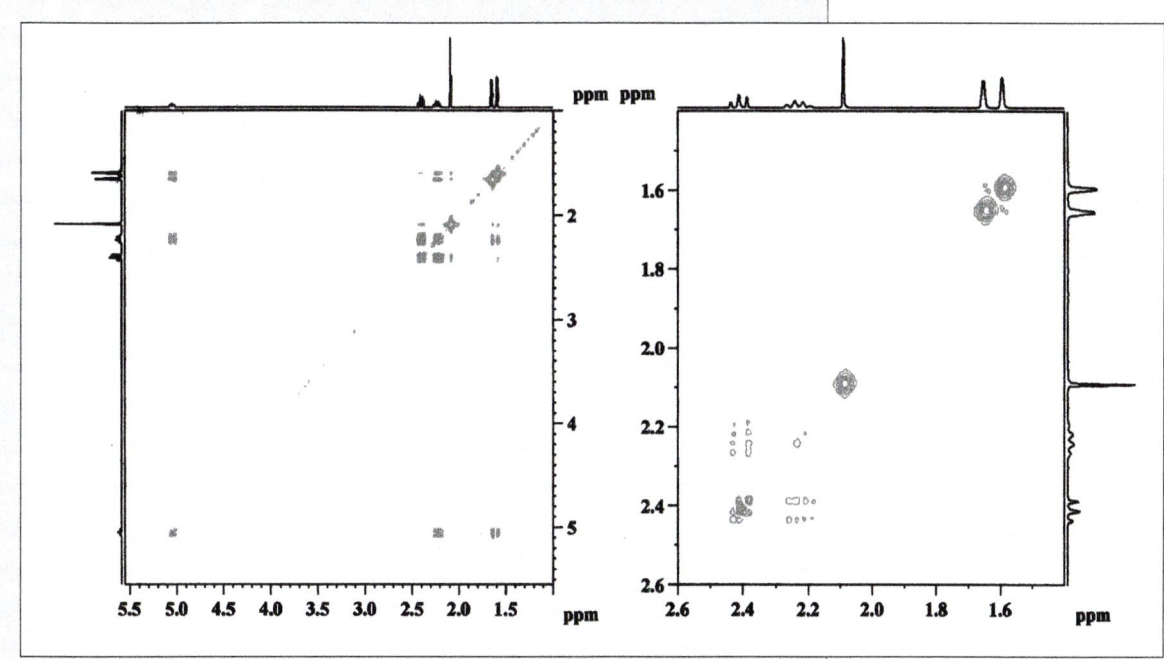

Figure 6.18   300 MHz HH COSY NMR spectrum of unknown **C** in CDCl₃

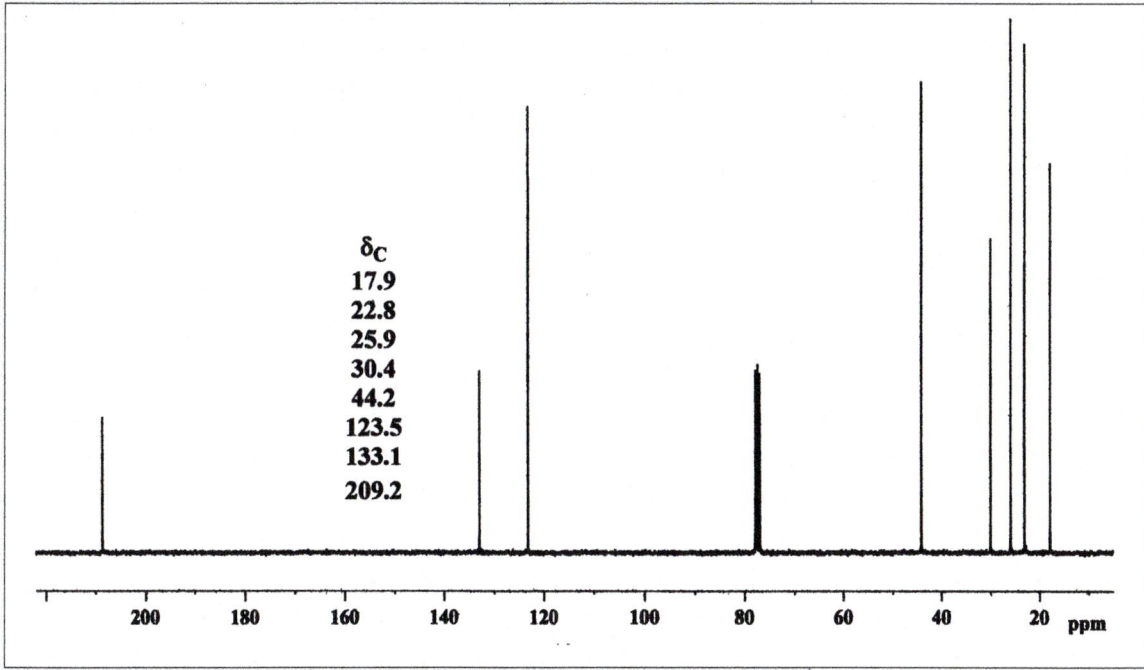

δC
17.9
22.8
25.9
30.4
44.2
123.5
133.1
209.2

Figure 6.19   75 MHz proton-decoupled ¹³C NMR spectrum of unknown **C** in CDCl₃
(δC 77.6)

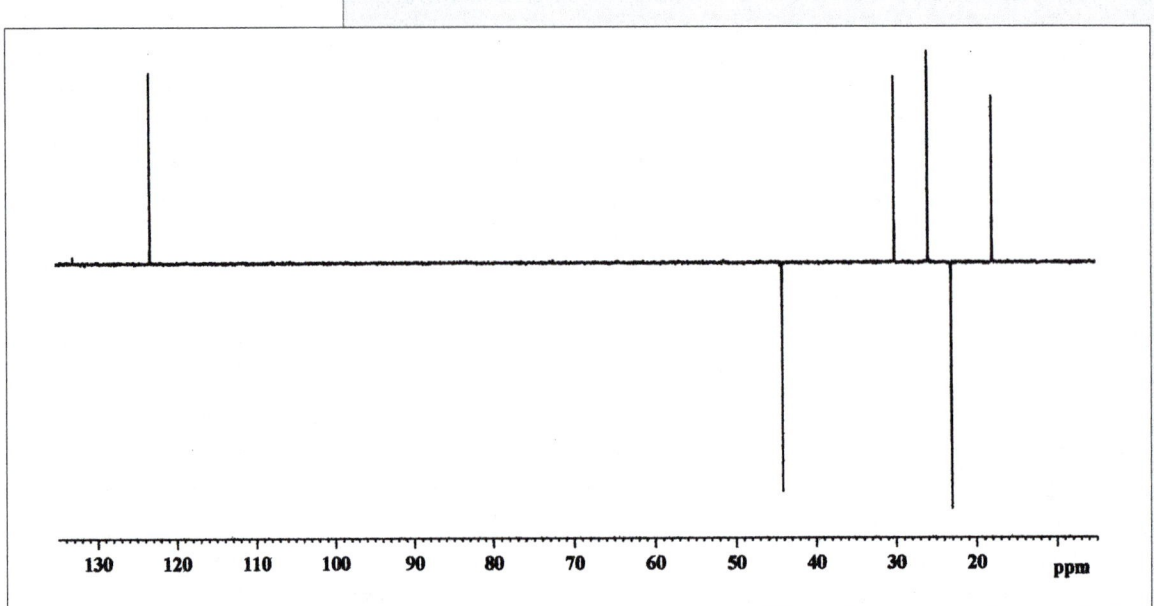

Figure 6.20   75 MHz DEPT 135 NMR spectrum of unknown **C** in CDCl₃

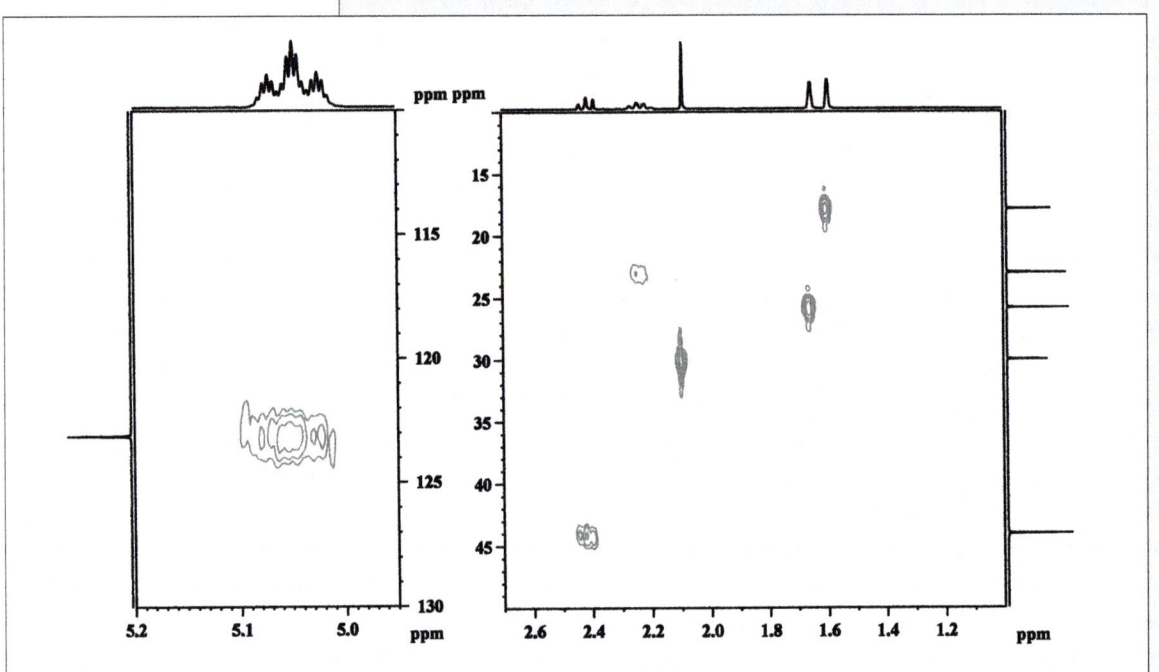

Figure 6.21   HMQC [¹*J*(¹H–¹³C)] NMR spectrum of unknown **C** in CDCl₃

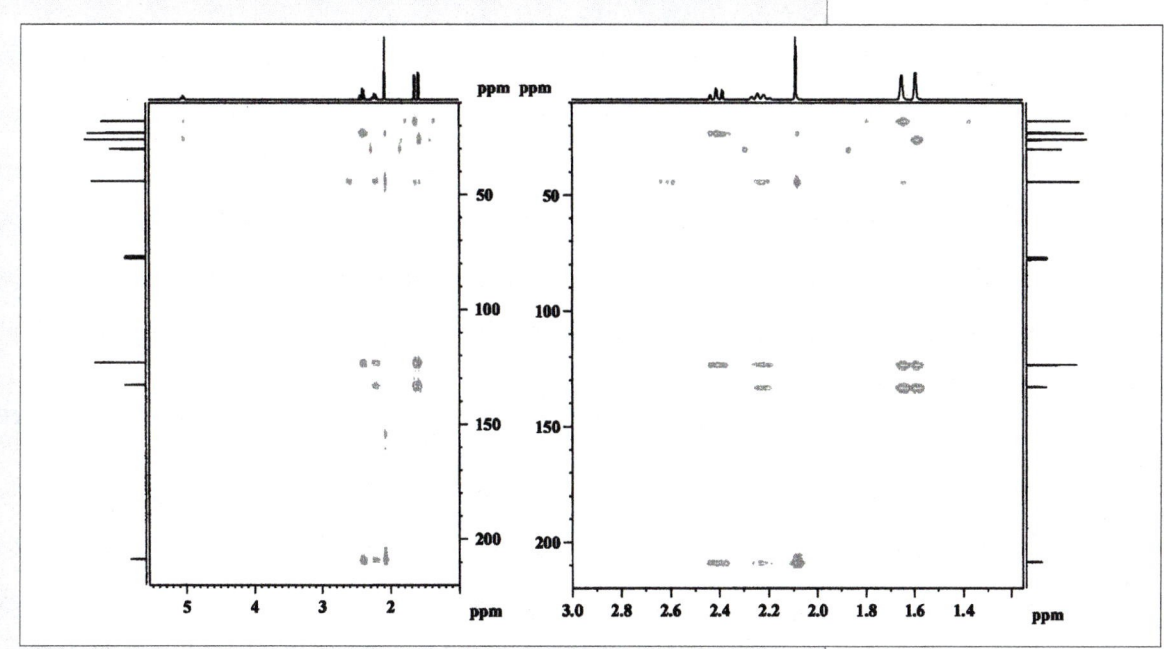

Figure 6.22   HMBC [$^2J(^1\text{H}-^{13}\text{C})$ and $^3J(^1\text{H}-^{13}\text{C})$] NMR spectrum of unknown **C** in CDCl$_3$

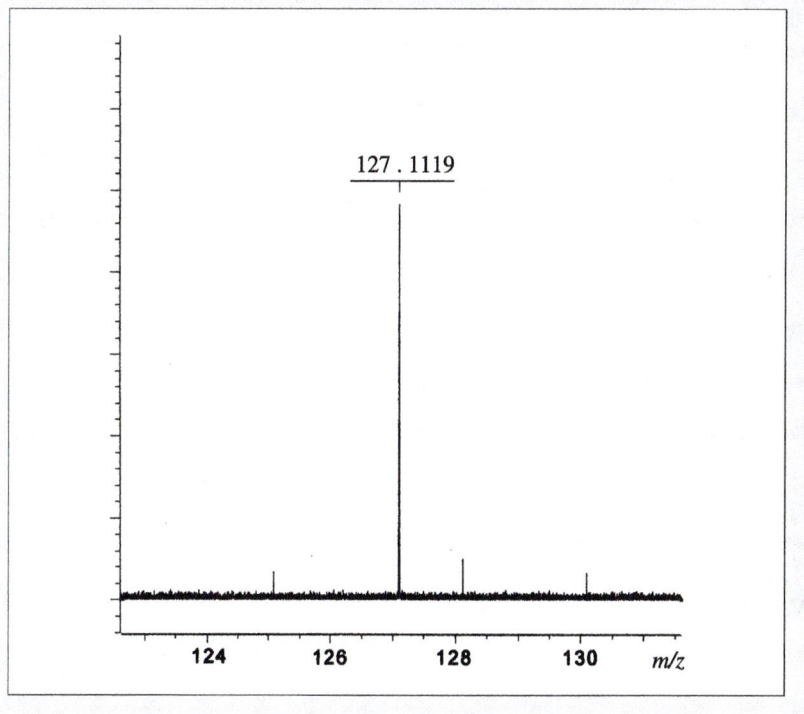

127 . 1119

Figure 6.23   Accurate mass APCI+ spectrum of unknown **C**

Further Reading and Useful Websites

General

D. Williams and I. Fleming, *Spectroscopic Methods in Organic Chemistry*, 5th edn., McGraw-Hill, New York, 1995.

NMR

T. D. W. Claridge, *High-Resolution NMR Techniques in Organic Chemistry*, Pergamon Press, Oxford, 1999.
J. K. M. Saunders and B. K. Hunter, *Modern NMR Spectroscopy, A Guide for Chemists*, 2nd edn., Oxford University Press, Oxford, 1993.
E. Breitmaier, *Structure Elucidation by NMR in Organic Chemistry*, Wiley, Chichester, 1993.

IR

K. Nakanishi and P. H. Solomon, *Infrared Absorption Spectroscopy*, 2nd edn., Holden-Day, San Francisco, 1977.

Websites

SDBSWeb: http://www.aist.go.jp/RIODB/SDBS/menu-e.html
WebSpectra: http://www.chem.ucla.edu/~webspectra/

Answers to Problems

Chapter 1

1.1. (a) 1.99×10^{-25} J (NMR); (b) 7.83×10^{-19} J (UV-Vis); (c) 1.99×10^{-23} J (microwave); (d) 5.17×10^{-20} J (IR); (e) 2.72×10^{-19} J (UV-Vis); (f) 3.99×10^{-17} J (X-ray).

1.2. (a) 1; (b) 5; (c) 5; (d) 6; (e) 4; (f) 20.

Chapter 2

2.1. The lone pair on the nitrogen atom of aniline is conjugated with the aromatic ring, which increases the chromophore of the aromatic ring from 203 nm for benzene to 230 nm for aniline. When an electron-withdrawing group, such as a nitro group, is added to aniline in the 2- (*ortho*) or 4- (*para*) position, the π-system of the aromatic ring is extended from the nitrogen of the amine to the nitro group:

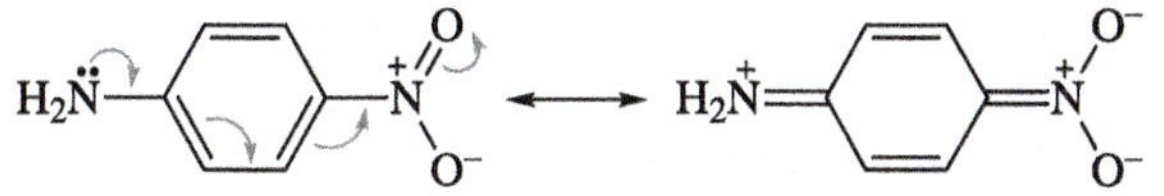

This leads to a decrease in the energy difference between the HOMO and LUMO orbitals and an associated red or bathochromic shift (to longer wavelength), along with an increase in the intensity of the absorption.

2.2. The calculated and observed absorbance wavelengths for both carvone and piperonal are in good agreement:

Carvone	Calculated λ_{max}		nm
	Parent enone		215
	(X = C, six-membered ring)		
	R groups	α	10
		β	12
	Total		$\overline{237}$
	Observed λ_{max}		236
Piperonal	Calculated λ_{max}		nm
	Parent aromatic carbonyl		250
	(X = H)		
	Substituents	*meta*	7
		para	25
	Total		$\overline{282}$
	Observed λ_{max}		280

The energy of an absorbance can be calculated from the wavelength using equation (2.2):

$$E(\text{kJ mol}^{-1}) = \frac{1.19 \times 10^5}{\lambda(\text{nm})}$$

For carvone, $\lambda = 236$ nm:

$$\therefore E\,(\text{kJ mol}^{-1}) = \frac{1.19 \times 10^5}{236} = 504.2\,\text{kJ mol}^{-1}$$

For piperonal, $\lambda = 280$ nm:

$$\therefore E\,(\text{kJ mol}^{-1}) = \frac{1.19 \times 10^5}{280} = 425.0\,\text{kJ mol}^{-1}$$

Thus piperonal requires the least energy (for the transition at 280 nm).

2.3. The molar absorptivity (ε) can be calculated from the data provided, which will give ε in units of $\text{dm}^3\,\text{mol}^{-1}\,\text{cm}^{-1}$, using equation (2.3):

$$A = \varepsilon c l$$

The equation requires rearrangement to:

$$\varepsilon = \frac{A}{cl} \text{ and for both of these examples, } l = 1 \text{ cm.}$$

For carvone, $A = 0.432$ and $c = 0.043 \times 10^{-3}$ mol dm^{-3}:

$$\therefore \varepsilon = \frac{0.432}{0.043 \times 10^{-3}} = 10,046 \, \text{dm}^3 \, \text{mol}^{-1} \, \text{cm}^{-1}$$

For piperonal, $A = 0.509$ and $c = 0.067 \times 10^{-3}$ mol dm^{-3}:

$$\therefore \varepsilon = \frac{0.509}{0.067 \times 10^{-3}} = 7597 \, \text{dm}^3 \, \text{mol}^{-1} \, \text{cm}^{-1}$$

Chapter 3

3.1. Unknown A: 5 DBE; –NH$_2$, C=O, OH. Structure:

3.2. Unknown B: 6 DBE; –NHCO–, CO$_2$H, C=C. Structure:

3.3. Unknown C: –NHCO–, C=O, C=C. Structure:

3.4. **E** is Figure 3.25 and **D** is Figure 3.26.

3.5.

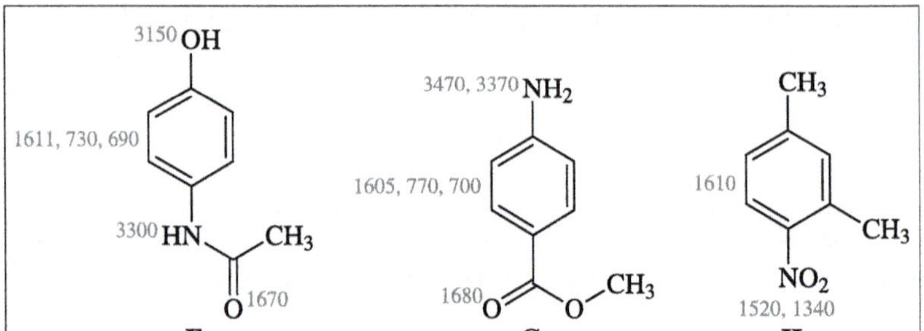

F, G, H structures

Chapter 4

4.1. **A** is 2-phenylpropane and **B** is 1-phenylpropane:

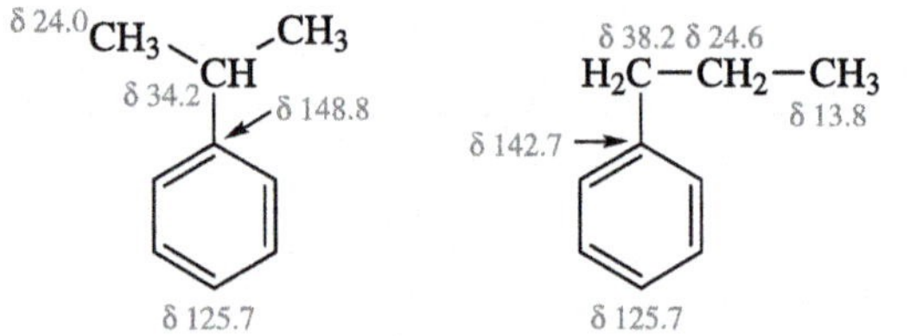

A, B structures

δ_C for alkylation products: assignment of the CH signals at δ 126.4 and 128.3 in **A** and 128.3 and 128.5 in **B** would require the use of HMBC spectra.

4.2. The δ values for **C** are assigned as shown:

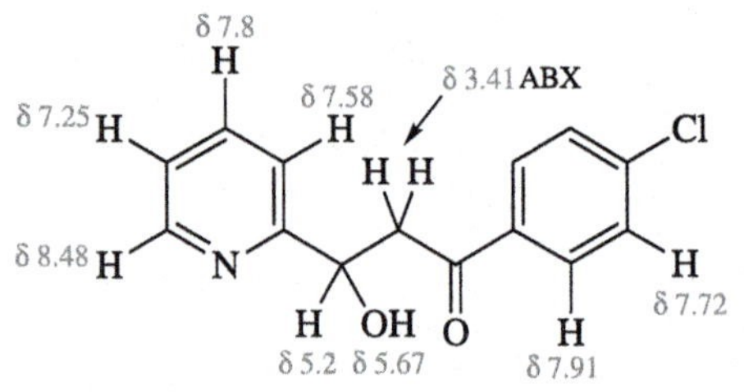

C structure

4.3. The oxidation has been a success. The introduction of a double bond between C2 and C3 has decreased the distance between H^3 and the CH_3 protons and they now undergo ^4J coupling ($J = 6.6$ Hz):

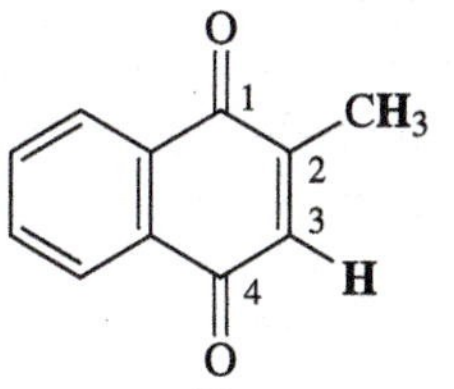

D

4.4. See Figure 4.20.

4.5. The coupling constants are:

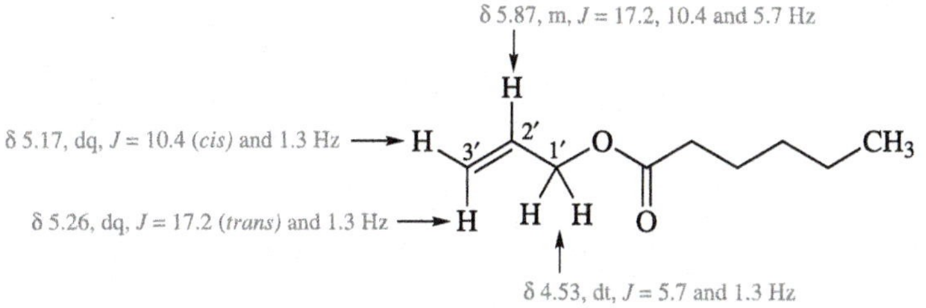

δ 5.87, m, $J = 17.2$, 10.4 and 5.7 Hz

δ 5.17, dq, $J = 10.4$ (*cis*) and 1.3 Hz

δ 5.26, dq, $J = 17.2$ (*trans*) and 1.3 Hz

δ 4.53, dt, $J = 5.7$ and 1.3 Hz

Note that, in this case, $^4J(H^{1'}-H^{3'}) = {}^3J(H^{1'}-H^{2'})$

4.6. Assignment: δ_H 3.17 (1H, dd, $J = 9.2$, 8.0, H2), 3.29 (1H, dd, $J = 9.8$, 9.2, H4), 3.37 (1H, ddd, $J = 9.8$, 6.0, 2.3, H5), 3.40 (1H, t, $J = 9.2$, H3), 3.49 (3H, s, OCH_3), 3.63 (1H, dd, $J = 12.3$, 6.0, H6a), 3.84 (1H, dd, $J = 12.3$, 2.3, H6b), 4.29 (1H, d, $J = 8.0$, H1):

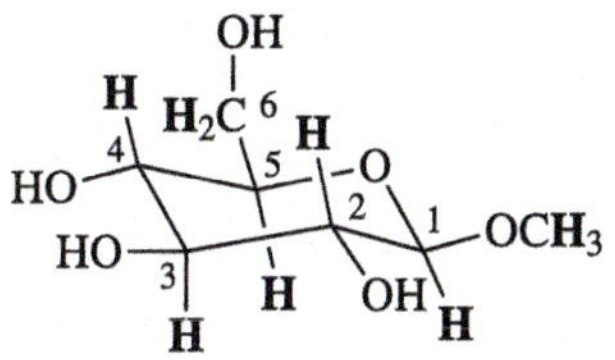

Chapter 5

5.1. By decreasing the potential difference the electrons which bombard the sample are accelerated through, so that they have an energy of less than 70 eV (for example, 20 eV) and their impact leads to less fragmentation of the sample.

5.2. The acid formed from ammonia (NH_4^+) is not as strong as that from methane (CH_5^+) so that, in cases where the sample is ionized by the ammonium ion (NH_4^+), this process is less energetically favourable than the corresponding process with the carbonium ion (CH_5^+) and therefore less fragmentation usually accompanies the use of ammonia as the reagent gas.

5.3. The structures of **A** and **B** and their fragmentation products are:

A 1-Phenylpropane m/z 91

B 2-Phenylpropane m/z 105

5.4. (a) m/z 609.2801 (100%); (b) m/z 610.2835 (37%).

5.5. The structures of the daughter ions and their mechanisms of formation are:

m/z 121 m/z 93
(formed *via* β-cleavage of C—O bond)

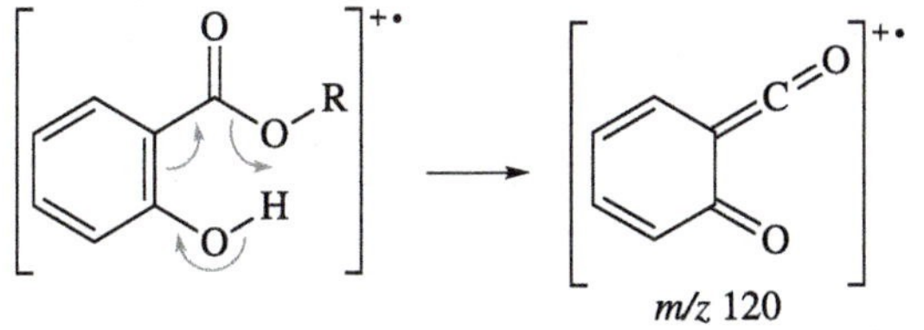

m/z 120

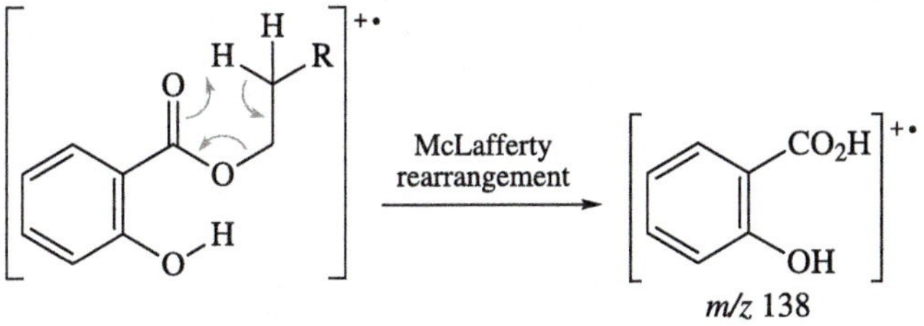

McLafferty rearrangement

m/z 138

5.6. The chloride counter-ion does not show up in the ESI + spectrum.

$$m/z\,306 : 2 \times {}^{35}\text{Cl}, 17 \times {}^{12}\text{C}$$
$$m/z\,307 : 2 \times {}^{35}\text{Cl}, 16 \times {}^{12}\text{C}, 1 \times {}^{13}\text{C}$$
$$m/z\,308 : {}^{35}\text{Cl}, {}^{37}\text{Cl}, 17 \times {}^{12}\text{C}$$
$$m/z\,309 : {}^{35}\text{Cl}, {}^{37}\text{Cl}, 16 \times {}^{12}\text{C}, 1 \times {}^{13}\text{C}$$
$$m/z\,310 : 2 \times {}^{37}\text{Cl}, 17 \times {}^{12}\text{C}$$
$$m/z\,311 : 2 \times {}^{37}\text{Cl}, 16 \times {}^{12}\text{C}, 1 \times {}^{13}\text{C}$$

Chapter 6

6.1. The expected accurate mass of **B** (MH^{+}, $C_8H_7O_3$) is 151.0393. The structure is:

B $C_8H_6O_3$, DBE = 6

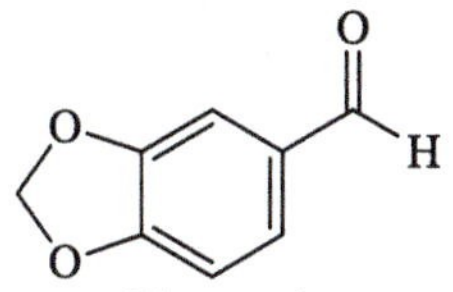

Piperonal

6.2. The expected accurate mass of **C** (MH^+, $C_8H_{15}O$) is 127.1119. The structure is:

C $C_8H_{14}O$, DBE = 2

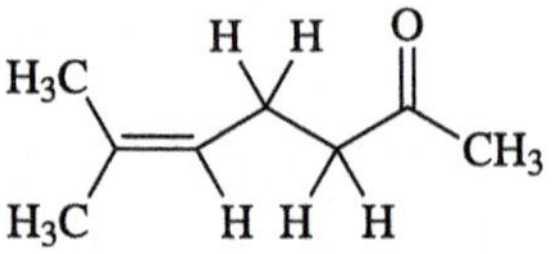

6-Methylhept-5-en-2-one

Subject Index